THE PROBLEM OF
THE RATIONAL SOUL IN THE
THIRTEENTH CENTURY

BRILL'S STUDIES IN INTELLECTUAL HISTORY

THE PROBLEM OF
THE RATIONAL SOUL IN THE
THIRTEENTH CENTURY

BY

RICHARD C. DALES

E.J. BRILL
LEIDEN · NEW YORK · KÖLN
1995

The paper in this book meets the guidelines for permanence and durability of the Committee on Production Guidelines for Book Longevity of the Council on Library Resources.

B
738
.S68
D35
1995

Library of Congress Cataloging-in-Publication Data

The CIP-data has been applied for.

Die Deutsche Bibliothek - CIP-Einheitsaufnahme

Dales, Richard C.:
The problem of the rational soul in the thirteenth century / by Richard C. Dales. – Leiden ; New York ; Köln : Brill, 1995
  (Brill's studies in intellectual history ; Vol. 65)
  ISBN 90–04–10296-5
NE: GT

ISSN 0920-8607
ISBN 90 04 10296 5

PRINTED IN THE NETHERLANDS

CONTENTS

CONTENTS

PREFACE

Over the past twenty-five years, I have been curious about what lay behind the Parisian condemnation of March 7, 1277, and I have published a number of articles, books, and editions concerning the animation of the heavens, the doctrine of the double truth, and the eternity of the world. I have avoided the study of the rational soul because of its complexity and difficulty. Fortunately, during that period a great deal of work was done on the subject by many excellent scholars, especially the identification and edition of a number of texts. This has simplified my task greatly. Still, the reader will notice that I do not always agree with my predecessors' understanding of the doctrine of some masters; or perhaps more accurately, I have been impressed by different aspects of some masters' thought than they have. But I have never departed from a received interpretation without careful thought. Some scholastics are particularly difficult to understand on certain points. Those who have given me the most trouble are Philip the Chancellor, Peter of Spain, and Albert the Great. I hope that I have finally got them right.

Most of those who have worked on this subject have understandably been either philosophers or theologians. I am neither; I am a historian, and what I have set out to write is a history of how thinkers dealt with the problem of the rational soul from the time of Gundissalinus to the 1277 condemnation. My ending point is somewhat arbitrary, for interest in the human soul, unlike that concerning the eternity of the world, was constant throughout the Middle Ages and Renaissance. However, I found the period I had set for myself to be a sufficient challenge, and I hope it has some value in adding to our understanding of the subject.

I have made no attempt to be comprehensive. Instead I have chosen to discuss those authors whose works contributed to the story and are available in print. With few exceptions (all noted), I have translated all the Latin in the text itself and have provided the Latin in the notes..

I am particularly indebted to three friends, Prof. Joseph Goering, Prof. R. James Long, and Prof. Gary Macy, who have taken time out from their own work to read this book. They have all provided me with excellent criticism, which has considerably improved my original draft and saved me from some embarrassing blunders. But I did not take all the advice I received, and I remain solely responsible for the book as it stands.

CHAPTER ONE

THE PROBLEM

There have been some excellent studies of certain aspects of thought on the soul in the early and middle thirteenth century, especially those of J. M. Da Cruz Pontes, Roberto Zavalloni, and Bernardo Bazán, as well as many helpful studies of individual thinkers. In addition, quite a high proportion of the known texts has been edited, thus greatly facilitating a study of the subject. The 'heterodox Aristotelians' or 'Latin Averroists' have also received a great deal of attention during the past fifty years, so that it is now possible to make suitable distinctions among them and appreciate the rationale of their enterprise more fully than was possible a century ago. Most of the studies of the earlier scholastics ('pre-Thomists') have centered on the problem of the unity of substantial form or the hylomorphic composition of spiritual substances, since these problems were initially raised regarding the human soul. Although it is necessary to take these doctrines into consideration in any discussion of the soul in the thirteenth century, I am not, in this work, primarily concerned with classifying writers according to their views in these questions, or in determining whether they were Avicennists, Augustinians, or Aristotelians—indeed, they were all indebted to some degree to all three of these *auctoritates*—but rather in looking anew at the range of texts from Gundissalinus to the 1277 condemnation in order to appreciate better the entire tradition.

Since we shall encounter many instances in this study of a writer's stating expressly that man has only one substantial form, the rational soul, but giving an account that seems to imply something different, it would be well to repeat here the distinction between the two positions (i.e., the unity or plurality of substantial form) that Fr. Daniel Callus has stated with admirable clarity and precision:

> If with Aristotle one holds (i) that prime matter is a completely passive potency without any actuality of its own whatever; (ii) that privation is the disappearance of the previous form, and, consequently, has no part at all in the composition of the substance; and (iii) that substantial form is absolutely the *first* determining principle, which makes the thing to be what it is, the only root of actuality, unity, and perfection of the thing; then, consistent with his stated principles, the conclusion forced on us is that in one and the same individual there can be but one single substantial

form: other forms, that come after the first, are simply
accidental and not substantial forms. Since the thing is
already constituted in its own being, they cannot give
substantial being, but exclusively accidental or qualified
being; they do not confer upon the concrete thing its own
definite and specific kind of being, e.g., man, but only a
qualified or relative state of being, for example, of being
fair or dark, big or small, and the like.

On the other hand, if one contends (i) that primary matter
is not absolutely passive and potential, but possesses in
itself some actuality, no matter how incomplete or
imperfect it may be: an *incohatio formae*, or any active
power; (ii) that privation does not mean the complete
disappearance of the previous form, so that matter is not
stripped of all precedent forms in the process of becoming;
or (iii) that substantial form either meets with some
actuality in prime matter or does not determine the
composite wholly and entirely, but only partially; from all
this it will necessarily follow that there are in one and the
same individual plurality of forms.[1]

The problem of the unity of substantial form first arose among the scholastics
in connection with the human soul, the Augustinians holding to a theory of
unity and the Aristotelians, for the most part, to a theory of plurality. But
these arguments concerned only the soul and not the soul's relation to the
body.[2] The two were considered by nearly all authors before Aquinas to be
distinct substances. But one must make suitable distinctions among our
authors concerning the way they understood their authorities and not be
unduly influenced by verbal formulas. With regard to the unity (or plurality)
of substantial form, one should realize that the question does not seem to have
been considered central before Aquinas. All authors subscribed to some
version of what, in retrospect, we may designate as a theory of plurality of
substantial form, at least if one considers a compound form, even though
unified, to be 'plural.'

The situation is further complicated by the distinction between the unity or
plurality of forms in any creature, and that between the unity and simplicity of
the human soul, on the one hand (a view derived from Augustine), and the

---

[1] Daniel A. Callus, "The Origins of the Problem of the Unity of Form," *The Thomist*
24 (1961), 257-88, on p. 258.

[2] On this question, see especially Roberto Zavalloni, *Richard of Mediavilla et la
controverse sur la pluralité des formes* (Louvain, 1951) and D. A. Callus, "The Origins of the
Problem of the Unity of Form."

compound nature of the soul (vegetative, sensitive, and rational) on the other, with which the Aristotelians struggled. For those who held that the human soul was simple still considered it to be the form, act, or perfection of the body, and considered the body to be a separate substance from the soul, another complete substance, and so they had to admit a proper form for the body in addition to the soul,[3] even as they insisted that the two substances made truly one thing, namely a human being. In such a context the distinction between unity and plurality of forms has little meaning, and arguing about which label to apply to an individual author is of little value in helping to understand his thought.

The question of matter-form composition of all beings other than God can also lead to misunderstanding if we pay too much attention to formulas and not enough to what they mean for each thinker. Avicebron's *Fons vitae* was the *locus classicus* for the doctrine, but even those who explicitly cite the work often mean something quite different by it than did Avicebron.[4] All scholastics, and even Averroes, admitted that there was something corresponding to matter even in spiritual substances, so that they could be distinct from each other and so that souls could be considered as substances and could interact in some way with corporeal creation. Whether this is called 'spiritual matter' and then qualified to the point that it has little to do with the doctrine of the *Fons vitae*, or whether it is called the distinction (derived from Boethius's *De hebdomadibus*) between *quod est* and *quo est*, which are then made analogous to matter and form, is not so important as what specifically was intended by the author. Pecham, for example, insists on the hylomorphic composition of souls but specifically equates his meaning with what not only Augustine (to whom, not Avicebron, he attributes the doctrine), but also Averroes taught. Universal hylomorphism can be grafted onto Aristotelian thought, as it was by Roger Bacon and Matthew of Aquasparta, or onto Augustine's, as it was by Pecham and many others. And during the first half of the thirteenth century, unity of substantial form was attributed to Augustine and plurality to Aristotle.

The meaning of the terms agent intellect and possible intellect was variously understood. The whole matter was confused by the Augustinian-Avicennan doctrine of the two faces of the soul, the Avicennan doctrine of the Giver of Forms, and Augustine's references to the 'uncreated light' that alone provides

---

[3] See Bernardo Bazán, "Pluralisme de forms ou dualisme de substances? La pensée pré-thomiste touchant la nature de l'âme," *Revue philosophique de Louvain* 67 (1969), 30-73.

[4] For a study of some of the changes rung on this theme, see D. O. Lottin, "La composition hylémorphique des substances spirituelles. Les debuts de la controverse," *Revue néoscholastique de philosophie* 34 (1932) 21-39.

certainty, so that the agent intellect is sometimes equated with the higher function of the soul, which attends to things above it and receives knowledge directly from above, and the possible is equated with the practical lower knowledge of the sense world. There were many variations of this. There was probably more confusion and disagreement on the meanings of agent and possible intellect than on any other aspect of thought on the soul.[5]

From the beginning, the study of the soul had lain within the province of both the arts and theological faculties; and although some theologians on occasion specified that they were writing strictly as philosophers, and many artists recognized theological truths as limits to their discussions, it is of central importance to our investigation to realize that the aims, procedures, and authorities of the artists and theologians were quite different. The primary task of the artist was to expound the meaning of the text he was teaching, and, in disputed questions that went beyond this limited aim, to solve the question by acceptable philosophical means: that is, by reason and experience. Theologians, on the other hand, had the task of determining the truth of any proposition on the basis of Scripture and the Fathers, aided as far as possible by the resources of philosophy.

Boethius of Dacia chose the problem of the eternity of the world to illustrate his point that theology and natural philosophy do not contradict each other, because the natural philosopher is concerned only with what can occur naturally but does not deny that things which could not occur naturally could occur by virtue of a higher, supernatural, cause, namely God's will.[6] But whereas this rationale works quite well with the question of a beginningless world, its pertinence to the debate on the nature of the soul is not nearly so clear.

THE BACKGROUND

Prior to the third quarter of the twelfth century, Latin Christian thought on the soul had been most heavily influenced by Augustine and by the pseudo-Augustinian works *De fide ad Petrum*, and *De definitione recte fidei*, (*De ecclesiasticis dogmatibus*). This tradition held that the soul was an immortal spiritual substance, created for and infused into the body of each individual by

---

[5] For a sampling of this confusion, see Leonard J. Bowman, "The Development of the Doctrine of the Agent Intellect in the Franciscan School of the Thirteenth Century," *The Modern Schoolman* 50 (1973) 251-279.

[6] *Boetii de Dacia Tractatus de aeternitate mundi*, ed. Géza Sajó (Berlin, 1964) and the discussion in R. C. Dales, *Medieval Discussions of the Eternity of the World* (Leiden, 1990) 145-154.

God. It ruled the body in this life, but it was capable of an independent existence and would be eternally punished or blessed as its actions in this world merited. It was in the soul that the human personality resided, and Augustine had defined man as a soul using a body.[7] There was always an incipient dualism in this tradition: soul and body were separate substances, but each soul had its own body, for which it had been created and to which it would be restored at the final Resurrection. But it had never been adequately explained how these two disparate substances constituted a truly unified human being.

It was the acquisition of a number of Muslim, Jewish, and Greek works during the twelfth century that cast the subject in a new light and made evident a large number of problems that had not been considered before. By the time Aristotle's *De anima* became generally known in the early thirteenth century, much of its teaching was already familiar to Latin scholars through Semitic intermediaries, but it had acquired a good deal of Neoplatonic baggage in the process of transmission. The most important of these authors were the Christian Arab Costa ben Luca, the Spanish Jew Avicebron, and the great Persian polymath Avicenna.

COSTA BEN LUCA. The Christian Arabic author Costa ben Luca, who lived from 864 to 923 A. D., was an important physician, philosopher, and translator of Greek works into Arabic. His short treatise *De differentia spiritus et animae*, translated into Latin by John of Spain, is a philosophical medical treatise, utilizing works by Plato, Aristotle, Galen, Hippocrates, Theophrastus, and others. The author first defines spirit and tells us its function:

> In the human body there are two spirits. One is called vital;
> its nourishment or sustenance is air. It emanates from the
> heart, and from there it is sent by the pulses through the rest
> of the body, and it brings about life, pulse, and breath. The
> other, which is called animal ... operates on the brain itself.
> Its nourishment is the vital spirit. It emanates from the
> brain, in which it brings about thought, memory, and
> foresight, and from there it is sent through the nerves to the

---

[7] "Homo igitur, ut homini apparet, anima rationalis est, mortali atque terreno utens corpore." *De moribus ecclesiae* 1, 27, 52, *PL* 32:1332; and "Nam mihi videtur [animam] esse substantia quaedam rationis particeps, regendo corpori accomodata." *De quantitate animae* 13, 22, *PL* 32:1048.

other members of the body to bring about sense and motion.[8]

Although there are many different opinions about the soul, he continues, we can learn from Plato and Aristotle that it is an incorporeal substance and that it is the ultimate source of the body's life and motion. Plato defines it as an incorporeal substance that moves the body. Aristotle, in his *De anima*, says: "Anima est perfectio corporis naturalis, instrumentalis, potentialiter vitam habentis." ("The soul is the perfection of a natural instrumental body potentially having life.")

The differences between soul and spirit are: 1- spirit is corporeal, soul is incorporeal; 2- spirit is contained by the body, but the soul cannot be contained by the body; 3- when the spirit is separated from the body, it perishes, but although the operations of the soul perish with the body, the soul itself does not. The soul moves the body and furnishes it with sense and life through the mediation of spirit, and spirit then performs these operations without any further mediator. The soul moves the body and furnishes it with life, because it is the first and greater cause of life. Spirit is the proximate cause of life, sense, and motion, and soul is the more remote cause.

This treatise was widely read and cited and, along with the *Fons vitae*, was the principal source among the Latins for the notion that the soul could only be united to the body through a mean.

AVICEBRON. The book *Fons vitae*[9] by Solomon Bar Jehuda ibn Gabirol, known to the Latins as Avicebron, exerted an influence on thirteenth century thought far beyond its evident merits. Written engagingly in dialogue form, the book had two main themes: that creation is not the result of necessity but of the absolutely free divine will (it was this especially for which William of Auvergne admired him); and that God's creation from nothing was restricted to universal matter and universal form. All other things were 'formed' from

---

[8] "[I]n humano corpore sunt duo spiritus: unus qui vocatur vitalis, cuius nutrimentum vel sustentatio est aër et ejus emanatio est a corde, et inde mittitur per pulsus ad reliquum corpus et operatur vitam, pulsum atque anhelitum; et alter, qui ... dicitur animalis, qui operatur in ipso cerebro, cujus nutrimentum est spiritus vitalis; et ejus emanatio est a cerebro, et operatur in ipso cerebro cogitationem et memoriam atque providentiam, et ex eo mittitur per nervos ad cetera membra, ut operatur sensus atque motum." Costa-ben-Lucae *De differentia animae et spiritus liber* 2. *Excerpta e libro Alfredi Anglici De motu cordis. Item Costa-ben-Lucae De differentia animae et spiritus liber translatus a Johanne Hispalensi*, ed. C. S. Barach (Innsbruck, 1878), p. 130.

[9] Edited by Clemens Baeumker, *Avicebrolis (ibn Gebirol) Fons Vitae ex Arabico in Latinum translatus ab Iohanne Hispano et Dominico Gundissalino*, Beiträge zur Geschichte der Philosophie des Mittelalters 1, 2-4 (Münster i.W., 1892-95) 1-339.

this original substance. It is the second of these two themes with which we are concerned here.

Universal substance, according to Avicebron, although its essence is multiple and diverse, can be resolved into two principles in which it is sustained and has its being: universal matter and universal form. "These two are the root of everything, and from them is generated whatever exists."[10] Universal matter has the property of existing per se; it is of one essence, it supports diversity, and it gives to all things their existence and name,[11] and there is only one such matter for all things.[12] But in order for it to exist *in effectu* form is required to bring about diversity and multiplicity.[13] Universal form has the properties of subsisting in another (i.e., matter), perfecting the essence of that in which it is, and giving being to it.[14]

Spiritual beings in Avicebron's thought were individuated by matter, not only with respect to species, but also with respect to individuals. All corporeal things contained the 'form of corporeity' and a hierarchical arrangement of subsequent forms, each of which inhered in the preceding form, so that every corporeal individual consisted of a plurality of forms.

The doctrine of this work was known to virtually every Latin scholastic of the thirteenth century and was accepted, sometimes with modifications, by many. It is not clear how many actually read the *Fons vitae* itself; only five manuscripts of the work are extant. But most of its teaching was contained in the works of Gundissalinus, which were widely disseminated.

AVICENNA. The works of the eleventh-century Persian scholar, philosopher, and physician, Avicenna (Ibn Sina) were incompletely translated into Latin, but they were among the first works of Muslim philosophy to be available to the Latins, and their influence on scholastic thought throughout the thirteenth century was second only to that of Aristotle. It is his *Liber sextus naturalium*,

---

[10] "[H]aec duo sunt radix omnium et ex his generatum est quicquid est." *Fons vitae, ed. cit.*, p. 7.

[11] "[Materia est] per se existens, unius substantiae, sustinens diversitatem, dans omnibus essentiam suam et nomen." *Fons vitae, ed. cit.*, p. 13.

[12] "... non quaesiuimus nisi unam materiam omnium rerum." *Loc. cit.*

[13] "Non dicimus materiam habere esse nisi cum conferimus ei formam spiritualem, in se autem non habet esse, quod habet cum adiungitur ei forma; et hoc est esse in effectu. alioquin cum dicimus eam esse, non habet esse nisi in potentia." *Fons vitae, ed. cit.*, p. 16.

[14] "Attende similiter proprietates formae universalis, quae sunt scilicet subsistere in alio et perficere essentiam illius in quo est et dare ei esse." *Loc. cit.*

or *De anima*,[15] with which we are concerned here. According to Avicenna, the soul and body are separate substances. The human personality resides in the soul. His thought experiment of the 'floating man' was frequently repeated by Latin writers: i.e., if a man were created fully grown but deprived of all sensory experience, he could still affirm his existence, and this would be the existence of his soul. The soul is not mixed with the body, nor is it dependent on it. Its relation to the body is accidental rather than substantial, for if it were substantial, then the soul would perish with the body.

The human soul may be considered from two points of view: as it is related to the body, and as it is in itself. In relation to the body, it may be called its form or perfection, since it animates the body and rules it. Seen from this standpoint, the soul's relation to the body is functional or operational. But this does not tell us what the soul is in itself. In itself it is a simple rational substance, possessing a multitude of faculties. These faculties include all the vegetative and sensitive operations as well as the rational. The soul has a natural inclination to· minister to the body, by reason of which it is individuated and by means of which it actualizes itself, but it does not need the body in order to subsist after this has been achieved.

There are four divisions of the speculative power of the soul: the material intellect, the intellect *in habitu*, the intellect in act (*in effectu*), and the *intellectus accommodatus*, which is fit to receive illumination from above. The soul may also be considered as having two faces:[16] the active, directed toward the body and practical affairs; and the contemplative, directed toward the intelligible objects provided by the Giver of Forms.

Full intellection is only possible because of the fully abstracted objects of knowledge coming by way of emanation from the separate Agent Intellect, or Giver of Forms, the Intelligence of the tenth and lowest celestial sphere. When the human intellect is in a fitting condition to receive its illumination, the Agent Intellect, which is itself always in act, completes the process of abstraction and brings about genuine understanding.

It will be evident in the pages that follow that a significant fraction of thirteenth-century thought on the soul is immediately derived from Avicenna's *De anima*, and that this work also largely determined the way in which Aristotle was understood. It is also evident that much of Avicenna's doctrine was sufficiently similar to Augustine's that the two could, without much difficulty, be conflated.

---

[15] *Avicennae Liber De Anima seu Sextus De Naturalibus*, ed. S. van Riet (Leiden, 1968).

[16] For a study of a portion of the influence of this doctrine, see Jean Rohmer, "Sur la doctrine franciscaine des deux faces de l'âme," *AHDLMA* 2 (1927) 73-77.

As a result of these newly acquired works, Latin scholars were confronted by a host of new questions, which would occupy them for the next century. The ultimate source for the new approach to the soul was Aristotle's *De anima*, which underlay the works of the Muslim and Jewish authors whose writings were known in the west some time before Aristotle's own treatise.

ARISTOTLE. Aristotle's *De anima* is in some respects his most impressive work and in others his most frustratingly incoherent and incomplete. Throughout most of the treatise the soul is considered to be the form of the body, the two constituting a single living substance,[17] the first perfection of a natural organic body having the potentiality for life. This, his most general definition of the soul, would imply that the soul, being the formal correlative of the body's matter and acting through the body's organs, would perish when the composite perished. This was in fact the way Alexander of Aphrodisias had understood him. But in a number of other places he muddied this clear concept, which would have been clearly heretical and would not have constituted a problem. In Book 1, 1 (403a) he said that "if some action or passion of the soul is uniquely proper to it, it is possible that it might be separated."[18] One such possible action or passion is understanding, and in *De anima* 1, 4 (408b) he notes that "the intellect seems to be a substance that comes about in a thing and is not corrupted."[19] And after discussing the problem, he concludes that the intellect is a thing of greater dignity and is something divine and impassible.[20] In 3, 4 (429b) he says that "the sense faculty is not outside the body, but the intellect is separated."[21] And in the

---

[17] I cite the text of the *translatio nova* from *Averrois Cordubensis Commentarium Magnum in Aristotelis De Anima Libros*, ed. F. S. Crawford (Cambridge, Mass., 1953): "anima est prima perfectio corporis naturalis habentis vitam in potentia. Et est secundum quod est organicum (2, 1, 412a, Crawford, p. 136) ... Si igitur aliquid universale dicendum est in omni anima, dicemus quod est prima perfectio corporis naturalis organici." (2, 1, 412b, Crawford, p. 138).

[18] "...si aliqua actionum aut passionum anime sit propria sibi, possibile est ut sit abstracta." Crawford, p. 18.

[19] "Intellectus autem videtur esse substantia aliqua que fit in re et non corrumpitur." Crawford, p. 87.

[20] "[I]ntellectus autem dignius est ut sit aliquod divinum et aliquod impassibile." (408b), Crawford, p. 89.

[21] "Sentiens enim non est extra corpus; iste [i.e., intellectus] autem est abstractus." Crawford, p. 417.

difficult chapter 5 of book 3, in distinguishing the active and passive intellects, he writes:

> Therefore it is necessary that in [the soul] there be an intellect capable of becoming all things, and an intellect capable of making itself understand all things. And the intellect which is capable of understanding all things is like a condition, such as light, for light in a certain way makes potential colors be actual colors. And this intellect is separated, not mixed or passible, and, in its substance, is action. ... Nor does it sometimes understand and sometimes not. And in its separated state, it is just what it is, and this alone is always immortal. And there is no memory, because [the agent intellect] is not passible, and the passible intellect is corruptible, and without it nothing is understood.[22]

And in *De generatione animalium* 2, 3 (736a), after discussing how the matter supplied by the mother is formed by the vital heat supplied by the father so that first the vegetative soul, having existed potentially in the semen, comes into being actually, and the sensitive soul similarly comes into actual being after having existed potentially in the vegetative, Aristotle concludes that the intellective soul cannot have been generated internally. "It remains," he says, "that the intellect alone should come from without, and that it alone be divine."[23] It is little wonder that there were, and still are, widely differing interpretations of Aristotle's teaching on the soul.

A number of new questions arose as a result of the gradual penetration of these new works into the intellectual community of western Europe, and some old questions gained a new significance. One group of such questions concerned the relationship of the soul and the body: Was the soul an individual substance

---

[22] "Oportet igitur ut in ea sit intellectus qui est intellectus secundum quod efficitur omne, et intellectus qui est intellectus secundum quod facit ipsum intelligere omne, et intellectus secundum quod intelligit omne, quasi habitus, qui est quasi lux. Lux enim quoquo modo etiam facit colores qui sunt in potentia colores in actus. Et iste intellectus etiam est abstractus, non mixtus neque passibilis, et est in sua substantia actio. ... Neque quandoque intelligit et quandoque non intelligit. Et cum fuerit abstractus, est illud quod est tantum, et iste tantum est immortalis semper. Et non rememoratur, quia iste non est passibilis, et intellectus passibilis est corruptibilis, et sine hoc nichil intelligitur." Crawford, pp. 437, 440, 443.

[23] "Restat igitur ut mens sola extrinsecum accedat, eaque sola divina sit." *De generatione animalium* 2, 3 (736a). *Aristotelis Opera cum Averrois Commentariis* (Juntine Press: Venice, 1562-74, reprint Frankfurt a. M., 1962) VI, fol. 74D-H.

in its own right, a *hoc aliquid*?[24]   Nearly all authors said it was. But this created several problems when Aristotle's definition of the soul as the form, or perfection, of the body was encountered (this ambiguity, as we have seen, existed even within Aristotle's *De anima*). First of all, is the soul the *first* perfection of the body? If the soul is an individual substance, how can it also be a form? And how do the body and soul, two separate substances, constitute one unified human being?

A second group of questions revolved around the soul's simplicity: Was it in fact simple or composed? If composed, does it therefore contain matter and form, and if it does, how can it in turn be a form? Or are there other things of which it might be composed?

A third group had to do with the origin of the soul: Was it *ex traduce*, that is passed along by the parents to the child by biological means? Two of the greatest of the Fathers, Augustine and Gregory the Great, had had serious doubts about this question. If it is not (and nearly all authors said that it was not), how was the rational soul related to the vegetative and sensitive souls, which clearly seem to be *ex traduce*? Is the rational soul a different kind of thing from the souls of beasts, or can there be two (or perhaps three) souls in one human being? If the human soul is a composite one, partly *ex traduce* and corruptible (the vegetative and sensitive parts) and partly *de novo* (the rational), how can it be said that the human soul is immortal? If one holds with Augustine that the rational soul is the only soul of a human being and performs all the vital functions, how does one account for the development of the embryo prior to the infusion of the rational soul?

Another group had to do largely with terminology and with the interpretation of Aristotle's words: What is the agent intellect and how is it related to the human rational soul? Is the possible intellect unique for the human species, or is it "multiplied according to the number of men"? What did Aristotle mean when he said that the intellect was separated, unmixed, and from without? Did he only mean that it operated without a bodily organ, or did he mean that it was unique like a separated celestial substance? It would seem that if it were immaterial it would have to be unique, and if, on the other hand, it contained matter and were a *hoc aliquid* it would be incapable of being the form of another substance.

Related to these, but not so essential for our investigation, were various explanations of the process of intellection, ranging from Augustinian-Platonic innatism to the Aristotelian *tabula rasa* account of the possible intellect.

It was largely this group of questions which occupied scholastic thinkers from the late twelfth century through the 1270s. As time went on, the

---

[24]   The term *hoc aliquid* (*lit.* "this something") is the usual Latin translation of Aristotle's τόδε τι (cf. *Metaph.* IV, 8, 1017b 10-24 and VIII, 7, 1049a 25-35).

problems were defined more precisely, but the suggested answers had much in common.

CHAPTER TWO

THE EARLY SCHOLASTICS

GUNDISSALINUS.   The first Latin treatise on the soul that was significantly influenced by Arabic treatments was the *De anima* of Gundissalinus,[1] written probably around 1170-75. This work was, in the words of Étienne Gilson, "a mosaic [composed] ... from small fragments borrowed here and there and juxtaposed in such a way as to form a whole."[2]   Gundissalinus's sources however were manifold and included Christian and Jewish authors, as well as Arabic, several of which Gundissalinus had had a part in translating.[3]   The most important sources were Avicenna's *De anima (Liber sextus naturalium)* and numerous writings of Augustine. The only portion of the treatise that cannot be traced to an earlier source is the mystical ending. Some scholars have considered this treatise to be of purely Arabic inspiration. But Gilson has shown, in an uncharacteristically testy essay, that, for all the Arabic influence, Gundissalinus's *De anima* is Christian in its essence, particularly in that it suppresses Avicenna's agent intellect and puts God in its place.[4] He is the first of a large number of Latin authors to do this. For all that, the treatise was considered flawed by at least one medieval reader who, at the end of the work in the Vatican manuscript, wrote: "Lege tractatum et fuge errores."[5]

In his Prologue, Gundissalinus sets forth his quintessentially scholastic goal, to enable the reader to understand the soul not only by faith, but also by reason. The soul, he says, is that which animates, "sensifies," and moves bodies by a voluntary motion. It is a substance, but not a body. He then gives a version of Aristotle's definition of the soul, which he took from Costa ben Luca's *De differentia spiritus et animae*: "Anima est prima perfectio corporis naturalis, instrumentalis, viventis potentialiter."

---

[1]   J. T. Muckle, C.S.B., ed., "The Treatise *De anima* of Dominicus Gundissalinus," with an introduction by Étienne Gilson, *Mediaeval Studies* 2 (1940), 23-102.

[2]   *Ibid.*, p. 24.

[3]   Specifically: Avicenna, *De anima* and *Metaphysica*; Algazel, *Summa theoricae philosophiae*; and Avicebron, *Fons vitae*. See M.-T. d'Alverny, "Dominic Gundisalvi (Gundissalinus)" in *New Catholic Encyclopedia* 4, pp. 966-67.

[4]   "The Treatise *De anima*," pp. 26-7.

[5]   Vat. lat. 2186, fol. 119v. See "The Treatise *De anima*," p. 28.

One of the questions that nearly all our authors will treat is one that had been the subject of a disagreement between Jerome and Augustine, namely whether the soul is *ex traduce*. After dicussing several views for and against, Gundissalinus leaves the question undecided but adds that "now however everyone holds that the soul is not *ex traduce*." New souls are created daily. According to some 'philosophers' they are created by angels rather than by God. Gundissalinus resists this but does admit that they are created by the 'ministerium' of angels, but by the authority of God. This seems to me to be a significant concession to Avicenna, even though Avicenna's "intelligentia" has been glossed "scilicet angelica creatura" by Gundissalinus. Gundissalinus also introduced from Avicebron's *Fons vitae* the doctrine that would later be known as the *binarium famosissimum*, that is, that everything other than God is composed of matter and form, whether corporeal or spiritual, with the consequence that there is a plurality of forms in any substance. The soul is simple, he says, although "it appears nevertheless to consist of matter and form. But this matter is incorporeal and spiritual, as is the soul's form. Composition of matter and form, he claims, does not destroy the soul's simplicity. Only God is absolutely simple. "No creature is absolutely simple but is called so only in comparison to other things."[6]

This simple soul performs all the function of the body. Since the vegetative and sensitive functions depend on bodily organs, and the soul can only exercise them through the body, their being is corporeal, and they do not survive the death of the body (although he contradicts this near the end of the treatise). Gundissalinus's treatment of the rational soul is a melding of elements from Avicenna's *De anima* and Augustine's *Soliloquia* and *De quantitate animae*. There are two powers of the rational soul, one of understanding, the other of acting. The first is the contemplative power, the other the active power, which are like the two faces of one substance, one of which looks downward to rule what is inferior to it (i.e., the body), the other of which gazes upward to contemplate that which is superior to it, which is God[7] (and significantly not the Intelligence of the tenth sphere, as in Avicenna). The former is the active intellect (*intellectus activus*, not to be confused with the *intellectus agens*, which Gundissalinus identifies with God), the latter the contemplative.

---

[6] "Solus ergo Deus absolute simplex est, et nulla creatura absolute simplex sed alterius comparatione dicitur." *De anima* 7, *ed. cit.*, p. 60.

[7] "Quae duae vires sive duo intellectus sunt animae rationali quasi duae facies; una quae respiciat deorsum ad regendum suum inferius quod est corpus ...; et aliam qua respiciat sursum ad contemplandum suum superius quod est Deus." *De anima* 10, *ed. cit.*, p. 84.

Through this work, Latin Europeans first became acquainted with a considerable portion of Jewish and Muslim interpretations of Aristotle's *De anima,* though in an ill-digested form. By mixing, or juxtaposing, elements from Augustine, Avicenna, and Avicebron, as well as by arbitrarily reinterpreting such notions as agent intellect, Intelligence, and so forth, Gundissalinus attempted to Christianize his exotic sources, but his attempt was only superficially successful. Still, for all its faults, this treatise was to be of seminal importance in subsequent Latin discussions of the soul. Father Daniel Callus has pointed out that it was Gundissalinus who introduced the Latins both to the universal hylomorphism of Avicebron, which underlay views of the plurality of substantial forms, and the doctrine of Avicenna, which underlay the view of the unity of substantial form. "The outstanding influence of this opuscule is amazing," he remarks, "in view of the mediocrity of its contents."[8]

ALEXANDER NEQUAM. During the last quarter of the twelfth century, the knowledge of the Arabic works increased greatly, and the text of Aristotle's *De anima* itself was gaining some currency. Alexander Nequam (1157-1217) studied and taught at both Paris and Oxford during this period, and after his retirement to the abbey of Cirencester in about 1200, he composed a diffuse compendium of theology (never completed) entitled *Speculum speculationum.*[9] It was written during the early years of the thirteenth century but was largely based an Nequam's notes from his Parisian sojourn from 1175-82 and his Oxford teaching during the 1190s. In addition to the Latin Fathers, he was quite well read in the recently translated Arabic literature, but he seems not to have known Aristotle's *De anima.*

In book 3 of the *Speculum speculationum* he discussed various problems concerning the soul. He had a good deal of trouble with the idea that beasts have souls, and he claims that he agrees with Aristotle (really Nicolaus of Damascus, *De plantis*) that they do not. The human soul, he says, is simple in its essence because it does not have integral parts. He denies that it is composed of matter and form ("although some people say that its matter is spiritual"), but then he goes on to assert that it is not simple in the same sense that God is, because it passed from non-being to being.

Then Nequam quotes Aristotle's definition of the soul from the *De anima* of his younger contemporary John Blund, but he is horrified by it: "*Anima est*

---

[8] D. A. Callus, "Gundissalinus' *De anima* and the Problem of Substantial Form," *The New Scholasticism* 13 (1939), 338-55, on p. 339.

[9] Rodney M. Thomson, ed., *Alexander Nequam Speculum Speculationum.* Auctores Britannici Medii Aevi 11 (London, 1988).

corporis organici perfectio, vitam habentis in potentia.   For since every
perfection is from form, it seems, according to this [definition], that the soul is
a form. What? Nay, every soul is a substance, but whether it is *ex traduce* or
not caused even Augustine to hesitate for a long time."[10]   Nequam took the
problem of the soul's origin very seriously and discussed it at length.   "I
cannot be persuaded that the soul is simple if it is *ex traduce*, he says, and he
concludes that it is not.

JOHN BLUND. Although a very learned man, Nequam was clearly not one of
the great intellects of his day. A much more impressive and competent work
on the soul was the *De anima* of John Blund,[11] written shortly after the year
1200, but whether at Oxford or Paris is uncertain. Blund, as was appropriate
for an artist, was particularly interested in the vegetative and sensitive
functions of the soul. He was well read in the voluminous twelfth-century
additions to the Latin scientific corpus, up to and including his older
contemporary Alfred Sareshel. He knows the *Philosophia* of William of
Conches, a number of Salernitan questions, Constantine the African's
Pantegni, and Isaac Israeli's *De dietis universalibus et particularibus*, to
name only a few. But he also asks and attempts to answer a number of
important basic questions.
   He is very clear that the soul is the form of the body, and he quotes
Aristotle's definition in a slightly different wording from that of any known
translation. "Whatever it is that moves the body by will," he says, "is the
soul." This something is "the perfection of an organic body, which possesses
the potentiality for life (*"Anima est corporis organici perfectio, vitam
habentis in potentia"*).[12]   Blund sees the problem implied by the soul's being
a form. He states it in an objection, and he attempts to solve it.

> Form confers being, and matter by itself is incomplete.
> Whence, every perfection is from form. Therefore, since
> the perfection of a body having organs possessing the
> potentiality for life would be the soul, the soul is a form.
> But no form exists by itself as a substance separate from

---

[10]   " 'Anima est corporis organici perfectio, uitam habentis in potentia.' Cum enim
omnis perfectio sit ex forma, uidetur secundum hoc quod anima sit forma. Quid? Immo
omnis anima substantia est, sed utrum ex traduce sit necne diutius hesitatus est eciam ab
Augustino." *Speculum speculationum* III, lxxxix, 1, *ed. cit.*, p. 359.

[11]   D. A. Callus O.P. and R. W. Hunt, ed., *Iohannes Blund Tractatus De Anima*.
Auctores Britannici Medii Aevi 2 (London, 1970).

[12]   Blund, *De anima* II, i, 14, *ed. cit.*, p. 5.

> matter. Therefore, the soul cannot be separated from the
> body, but perishes with the body.[13]

Gundissalinus had seen no inconsistency in asserting both that the soul is the form of the body and that it is composed of matter and form. Nequam had seen the problem and denied that the soul was a form. But Blund both sees the problem and, with the help of Avicenna, attempts to solve it: "The word soul ... signifies a kind of substance that is a kind of accident in relation to the organic body, insofar as it animates and vivifies that body by itself, and because of this accidental relationship it is said to be its perfection, that is, because it gives the body life."[14]

Blund dismissed the possibility of the soul's being *ex traduce* in a single sentence. But he undertook a lengthy investigation of the question of whether there are three souls in man or only one. He insists unequivocally that there is only one, but he presents an extensive account of arguments for the opposite position. His own solution is an ingenious attempt to differentiate the soul by logical relations, which will re-appear in about 1250 in Peter of Spain's *Scientia libri de anima*, during the 1250s in the works of Anonymous Admontensis and Anonymous Vaticanus, and in 1276 among the arguments in Question 22 of Siger of Brabant's commentary on the *Liber de causis*:

> The word soul signifies the genus of vegetable, sensitive,
> and rational soul; and in man there is one soul, which is
> vegetation, sense, and reason.   The sensible soul is a
> subaltern genus, because the sensible soul is the genus of
> the rational soul and the species of the vegetable soul. ... It
> is an incorporeal and incorruptible substance, because
> every universal is incorruptible.   Nevertheless, each soul,
> such as the vegetable soul that is in trees and the sensitive
> soul that is in beasts, is not incorruptible, for these can be
> corrupted according to their special being, but not
> according to their general being, by virtue of which they are
> incorporeal substances.   And this happens because their

---

[13] "Sed obicitur. Forma dat esse, et materia in se est imperfecta: unde omnis perfectio est a forma. Ergo cum perfectio corporis organici habentis vitam in potentia sit anima, anima est forma. Sed nulla forma est res per se existens separata a substantia. Ergo cum anima sit forma, anima non habet dici res per se existens separata a substantia. Ergo anima non potest separari a corpore, sed perit cum corpore." *De anima* II, i, 15, *ed. cit.*, p. 5.

[14] "... hoc nomen 'anima' ... [s]ignificat ... substantiam sub quodam accidente in relatione ad corpus organicum in quantum ipsum animatur et vivificatur per ipsam, et gratia illius accidentis dicitur esse perfectio ipsius, eo scilicet quod ipsa ipsum animat." *De anima* II, i, 16, *ed. cit.*, pp. 5-6.

special being can only be preserved in an organic body.
Whence, if the organic body is destroyed, both its being and
the being of its vegetable or sensitive soul, which is simple,
is also destroyed.[15]

So far, Blund has been talking about only the vegetable and sensible souls.
But what of the rational soul, which is the form of man, even though in an
'accidental' way? Unlike the vegetable and sensible, he says, it is neither a
body nor is it indebted to the nature of the body.

> The soul has twofold being: one from its first perfection,
> that is, from its first creation, and that is immortal and
> cannot be taken from it. Its second being is from the
> acquisition of knowledge and virtues; and it can be
> deprived of these if it should turn away from the way of
> pure truth and follow the path of falsehood. But its being
> ought not to be called mortal for this reason, but rather
> commutable and passible.[16]

Although the soul should be considered simple, since the First Cause is utterly
simple, and the things which come forth from it immediately (souls and
Intelligences) must have simple being, there is a kind of composition in it,
though not a composition of matter and form. "In the creation of the soul, its
components, i.e., the genus of soul and its differentia, did not exist separately
before the creation of the soul. And also, although there may be spiritual

---

[15]   "Dicimus quod hoc nomen 'anima' significat genus anime vegetabilis et anime
sensibilis et rationalis; et in homine est una sola anima a qua est vegetatio, sensus, et ratio. Et
anima sensibilis est genus subalternum, quia anima sensibilis est genus anime rationalis et
species vegetabilis. ... [E]st substantia incorporea et incorruptibilis, quia omne universale est
incorruptibile. Nec tamen quelibet anima est incorruptibilis, ut est anima vegetabilis que est
in arboribus, et anima sensibilis que est in brutis: hee enim secundum suum esse speciale
corrumpi possunt, sed non secundum suum generale esse secundum quod sent substantie
incorporee. Et hoc contingit quia suum esse speciale salvari non potest nisi in corpore
organico; unde destructo organico corpore destruitur et esse et esse ipsius anime, que simplex
est, vegetabilis vel sensibilis." *De anima* IV, 40-41, *ed. cit.*, p. 12.

[16]   "[A]nima duplex habet esse; unum a sua prima perfectione, scilicet a prima creatione, et illud est immortale et eo privari non potest; secundum eius esse est ab adquisitione
scientiarum et virtutum, et hoc esse potest ipsa privari, ut si ipsa puram veritatem respuat et
falsitatis viam intendat; sed propter hoc esse non debet ipsa dici mortalis, immo commutabilis
et passibilis." *De anima* XXIII, 324, *ed. cit.*, p. 88.

matter and spiritual form in the soul, it is not therefore corruptible, since it lacks the cause, contrariety, by which corruption takes place."[17]

The soul then is as simple as a created substance can be, but it is not utterly simple like God, since it has composition of genus and differentia (and it may contain spiritual matter, although Blund seems not to accept this). It is 'mortal' (commutable and passible) in its specific mode of existence, but it is immortal in its generic mode. Just this much of Blund's account might seem to permit the immortality only of the genus 'soul.' The rational soul, since it does not depend on the body, does not perish, even in its specific mode, with the death of the body. But Blund tells us nothing about how the genus 'soul' is individuated into specific souls. He is quite clear, however, that souls are individual.

His account of the process of intellection is, like Gundissalinus's, heavily dependent on Avicenna, but it is more clearly presented. It is the intellect's function to grasp both those things which are above human nature and natural things through the abstraction of universal forms from matter and from things appended to matter. It is reason's task to compare these intelligible things one with another and by comparing to make judgments about them. Avicenna distinguishes four modes of intellect: the material intellect, or intellect in potency; the formal or acquired intellect (*intellectus adeptus*); the intellect in effect; and the agent intellect.

The material intellect is the soul itself, bare of any acquired dispositions, and it is called material because it behaves in a manner similar to the aptitude of prime matter insofar as in itself it is not any form. The acquired intellect, which is the second perfection of the soul, is a passion generated in the soul, which is a similitude of a thing outside it. It is also called the formal intellect because it acts in a way similar to form, since, just as form perfects matter by giving it actual being, thus the acquired intellect perfects the material intellect by giving it being in effect. The intellect in effect is brought about by a conjunction of the material and acquired intellects, which actualizes its potency for intelligible form. And the agent intellect is the power of the soul that grasps universal things by abstracting them from accidents. As Blund understands the agent intellect then, it is a power of the individual soul rather than the Intelligence of the tenth sphere.

Although Blund was writing at the very beginning of the thirteenth century, he sees clearly what the basic problems are in trying to save both Aristotle's and Augustine's views of the soul, and his solutions, although far from

---

[17] "Non tamen ita fuit in creatione anime quod prius fuerunt illa componentia, scilicet genus anime et eius differentia ante creationem anime. Et etiam, licet in anima esset et materia spiritualis et forma spiritualis, non tamen esset ipsa corruptibils, cum causa contrariete careat secundum quam fit corruptio." *De anima* XXIV, 334, *ed. cit.*, p. 91.

satisfactory, are better than many of his successors would manage for two generations in the future. Although his understanding of Aristotle is still conditioned by the interpretations of Arabic intermediaries, Blund was better able to penetrate this exotic veil and get to the meaning of the text than scores of his successors, who had less excuse.

PHILIP THE CHANCELLOR. With Philip the Chancellor we encounter a scholar with a fairly wide, though not deep, knowledge of the Aristotelian corpus. He frequently cites the Aristotelian works of natural philosophy, including *De anima* (and even book 3, chapter 5), but he has not perceived their full import. He still depends heavily on Muslim interpreters, and his understanding of Aristotle, even when he had the text in front of him, seems to have been conditioned by his knowledge of Avicenna. Nevertheless, the Aristotelian concept of the soul and the problems arising from it have deranged the essentially Augustinian doctrine of the soul, which underlay much of the previous thought on the subject. Most of the questions he treats nevertheless arise from the works of Augustine, although their treatment is often influenced by Aristotle.

Philip wrote his *Summa de bono*[18] near the end of his life (probably *ca.* 1227-36), after a long career in the schools. But he kept his reading up to date. He not only quotes the works of his Latin contemporary Alfred of Sareshel, but his use of Maimonides and—more importantly for our purposes—Averroes are among the very earliest mentions of them in Latin Europe.[19] So although much of the *Summa de bono* was derived from Philip's actual teaching and disputations in the Parisian theological faculty, he did not simply copy out his earlier work as it fit into his scheme, but he continued to read and think.

Philip's position on whether or not the soul is composed of matter and form is ambiguous. In his question on the immortality of the soul, he presents a preliminary argument that definitely asserts the matter-form composition of the soul, but it is not clear what status he accords this argument:

---

[18] Leo W. Keeler, SJ, ed., *Ex Summa Philippi Cancellarii Quaestiones De Anima* (Münster, i.W., 1937) has been superseded by Nicolas Wicki, ed., *Philippi Cancellarii Parisiensis Summa de bono.* Corpus Philosophorum Medii Aevi. Opera Philosophica Selecta (2 vols., Bern, Switzerland, 1985).

[19] See R. A. Gauthier, "Notes sur les débuts (1225-1240) du premier 'Averroisme'," *Revue des scien-ces philosophiques et théologiques* 66 (1982), 322-74, which corrects R. de Vaux, "La première entrée d'Averroës chez les Latins," *Revue des sciences philosophiques et théologiques* 22 (1933), 193-245.

If one of a pair of correlatives exists, so does the other, and if one is able to be, so is the other. Form exists with respect to matter, and the active with respect to the passive. Therefore, if there is something in the universe whose property is a passive potency, such as first matter, which remaining uncorrupted receives all forms, it is necessary that its correlative be incorruptible. But the correlative of first matter is last form, and [the correlative of] passive potency [is] active potency. Therefore it is necessary for last form to have some active perfection and for this to be incorruptible. But the final perfection in nature is the rational soul. Therefore it is necessary that it be incorruptible.[20]

He treats the matter at greater length in his question "Quid sit anima:"

Next we must inquire whether the principles have matter and form, since they are not completely simple, as the First is. ... But in the soul there are two differences, one for receiving and another for acting. From this it is accepted that it has matter and form, since matter is the principle of receiving and form the principle of acting. For these two differences are the agent intellect and the possible intellect. Similarly, the same conclusion is reached from the fact that every being is simple or composite; and the soul is not simple, because then *quod est* and *quo est* would be the same thing in the soul. Furthermore, in the book *De trinitate* it is held that in everything caused by the First, there is "this" and "this," and thus matter and form. Furthermore, the same conclusion is drawn from the definition of the philosophers who have posited spiritual matter and spiritual form ... But there is in the rational soul a fuller approach toward composition of matter and form than in an Intelligence, which is shown in its nature to be united to a material body. Whence in the soul itself there

---

[20] "Correlativum si unum est, et reliquum, et si unum potest esse, et aliud; forma est respectu materie, et activum respectu passive. Ergo si aliquid est in universo cuius proprietas est potentia passive, sicut materia prima, que manens incorrupta suscipit omnes formas, necesse est eius correlativum esse incorruptibile. Sed prime materie opponitur ultima forma relative et potentie passive potentia activa. Necesse est ergo untimam formam habere perfectionem aliquam activam et eam esse incorruptibilem. Ultima autem perfectio in natura est anima rationalis. Quare necesse est eam esse incorruptibilem." *Summa de bono* IV, Q. 6, ed. Wicki, II, p. 267; ed. Keeler, p. 21.

> are the differences of agent and possible intellect, but there
> are not these differences in an angelic intelligence.[21]

In this question material has been taken from Boethius's *De trinitate*, Avicebron's *Fons vitae*, Alfred of Sareshel's *De motu cordis*, and Aristotle's *De anima* to show that the soul is not simple. It must be both receptive and active, it must be capable of union with the body, and one soul must be distinct from another. But whether its composition is of *quod est* and *quo est* only, as is the case with angels/Intelligences, or also of spiritual matter and form, is not clear; only that it has something analogous to matter and something analogous to form, and that the soul approaches matter-form composition more closely than do the angels because of its nature to be united to a body. He says that wherever there is matter, there is *id quod est*, but it is not necessary that wherever there is *id quod est*, there is matter. Philip's remarks on this question are not consistent. It would seem that he did not make a clear distinction between the *quod est* and *quo est* of Boethius and the spiritual matter and form of Avicebron.

Philip explicitly and consistently holds that the soul is both a substance and the form of the human being, but it is not only a form, "such as are other forms that are given by the Giver of Forms to prepared matter."[22]   Like Blund, he was aware of the difficulties involved in calling the soul the form of the body, while still considering it to be a substance: Form is united to matter *per se* and not by some other extrinsic principle, because they bespeak one and the same thing. "Therefore, 'that by which it is a being' and 'that by which they are one thing' are the same. But every being composed of matter

---

[21] "Deinde querendum est de principiis si habet materiam et formam, cum non sit omnino simplex quemadmodum Primum. ... in anima autem sunt due differentie, una ad recipiendum et altera ad faciendum. Ex quo accipitur quod habet materiam et formam, cum materia sit principium recipiendi, forma autem agendi; nam sunt eius hee differentie intellectus agens et possibilis. Similiter estimatur idem ex hoc quod omne ens simplex est aut compositum, et anima non est simplex, quia tunc idem esset in anima quod est et quo est ... Preterea, in libro De Trinitate habetur quod in omni causato a primo est hoc et hoc et ita materia et forma. Preterea, idem estimari potest ex diffinitione philosophorum qui posuerunt materiam spiritualem et formam spiritualem ... Est tamen in anima rationali amplior accessus ad compositionem materie et forme quam in intelligentia, quod ostenditur in natura eius secundum quam corpori material unitur. Unde in ipsa sunt differentie diverse intellectus et possibilis, in intellectu vero angelico non sunt its diverse differentie." *Summa de bono* IV, Q. 1, ed. Wicki, II, p. 157; ed. Keeler, p. 21. See also Lottin, "La composition hylémorphique," pp. 29-30.

[22] *Summa de bono* IV, Q. 6, ed. Wicki, II, p. 264; ed. Keeler, p. 51.

and form is a being through form. Therefore I say that it is one thing through the soul."[23]

But this is not the position he adopts. In responding to an argument against the soul and body being joined through a mean, Philip presents his view of a hierarchy of perfective and dispositive forms, each lower one acting as a 'material disposition' with respect to the next higher:

> The soul in one way is united in the manner of a form, and in another way in the manner of a substance. ... Some forms are first forms, some are last, and some are mediate. First forms, such as corporeity, since they are first, are joined to matter without a mean. Last forms are joined through a medium, and because they are last they are not means or natural dispositions for the conjunction of others; but the last of all natural forms is the soul. Mediate forms are sometimes joined by a mean and sometimes they behave as means or natural dispositions, as when the sensitive power is joined to its subject through a medium, namely the vegetative power, in those cases where the sensitive soul is the last perfection. But sometimes the very same thing, in a nobler subject, is a mean and, as it were, a material disposition, as the sensitive power is to the intellective soul. It is clear, therefore, that although it [i.e., the soul] behaves as a form, it is nevertheless not necessary that it be joined to the body *per se.* [24]

The same is true when we consider the soul as a substance. It must be united to the body through a mean because *per se* it is a substance separable from the body, and it is also separate in its operation. Therefore is does not depend on

---

[23] "Ergo idem est quo est ens et quo est unum. Sed omne ens compositum ex materia et forma est ens per formam; ergo per eam est unum, essentialiter dico." *Summa de bono* IV, Q. 8, ed. Wicki, II, p. 283; ed. Keeler, p. 81.

[24] "... anima secundum quid unitur per modum forme, secundum quid per modum substantie. ... [S]unt quedam forme prime, quedam ultime, quedam medie. Prime forme cum prime sint, absque medio materie coniunguntur, ut est corporeitas. Ultime forme per medium coniunguntur, et quia ultime non sunt media neque dispositiones materiales ad aliarum coniunctionem. Ultima autem forma naturalium est anima. Medie autem et per medium coniunguntur quandoque et quandoque sunt media et quasi materiales dispositiones; verbi gratia potentia sensibilis per medium coniungitur suo subiecto, scilicet mediante ut dispositione materiali potentia vegetabili; et hoc quando est ultima perfectio. Quandoque autem ipsa eadem in nobiliori subiecto est medium et quasi dispositio materialis, scilicet comparatione anime intellective. Manifestum est igitur quod, licet sit ut forma, non tamen per se corpori necesse est coniungi." *Summa de bono* IV, Q. 8, ed. Wicki, II, p. 284; ed. Keeler, p. 81.

the body according to either its substance or operation, and it requires some mean for its union with the body.[25]

The body itself is perfected and unified by the vegetative and sensitive powers, which are adaptive dispositions necessary for unifying the body and which are the media through which the soul and body are joined. The vegetative power is the remote medium, the sensitive the proximate. There are different types of bodies, whose generation requires varying degrees of complexity. Minerals require only a mixture of the elements; vegetables require in addition a complexion; and sensible beings require also a composition. Each is more perfect than the preceding simpler type. The highest type of creature, the rational, does not however arise through the operation of natural forces, but through creation. The rational soul and the body are opposed in three respects: the rational soul is simple, incorporeal, and incorruptible, whereas the body is composite, corporeal, and corruptible. The vegetative soul is composite, incorporeal, and corruptible; the sensitive is simple, incorporeal, and corruptible; the rational is simple, incorporeal, and incorruptible. So each level of soul shares two properties with those adjacent to it. At the extremes, only the body is corporeal, and only the rational soul is incorruptible. It is by these shared properties that they serve as the means for unifying the body and for uniting it to the rational soul. The sensitive soul is the perfection of the body,[26] but since it shares two characteristics, incorporeality and simplicity, with the rational soul, and one, corruptibility, with the vegetative, it can serve as the mean for uniting body and rational soul.

The resulting unity of the perfected human being is threefold. Philip distinguishes three degrees of unity. The first and highest degree is the completely inseparable unity of matter and form found in the supercelestial bodies. The second and lesser is found in all corruptible beings, in which matter is separable from form, but not form from matter. The third and least is found in those things where form and matter are joined so that one is separable from the other, as the soul from the body and the body from the soul. And so there is the least unity in man, who sums up (continues) all of creation from the lowest to the highest: "In a wonderful manner, man, since he continues higher and lower things in himself, has this threefold unity in himself, according to different aspects. For the first unity is in the soul itself

---

[25] *Loc. cit.*

[26] "... et est [anima vegetabilis] materialis dispositio ad recipiendum animam sensi-bilem in corpore, cuius est perfectio anima sensibilis." *Summa de bono* IV, Q. 8, ed. Wicki, II, p. 286; ed. Keeler, p. 86.

considered in itself; the second is in the body; and the third is in the whole man."[27]

This remarkable account of the compound soul and the threefold unity in man is also contained in the fragmentary anonymous *Summa Duacensis*, a work that is very similar to those parts of the *Summa de bono* with which it coincides. Victorin Doucet attributed the similarity to the fact that the *Summa Duacensis* was not a finished work prepared by the author for publication, but rather an early version of the material that Philip put into final form in the *Summa de bono*.[28]   Palémon Glorieux, however, the editor of the *Summa Duacensis*, has judged that there are too many differences of "style and presentation" for this to be the case and attributes it to an unknown master; but he also considers the *Summa Duacensis* to be the earlier work.[29]   Nicolas Wicki, the editor of the *Summa de bono*, presents the opinions of Doucet and Glorieux without attempting a solution, remarking that both schools of thought have found their champions.[30]   I personally think that Doucet's case is the more probable. But the hard evidence is slight and the arguments on both sides tenuous, and in any case it was through Philip's work that these views influenced generations of Latin thinkers.

The human soul then, according to Philip, is a compound soul. It is composed of three substances, two of which, the vegetative and the sensitive, are *ex traduce* and are corruptible, while the third, the rational, is created and immediately infused by God, and is incorruptible. These three substances, however, constitute only one soul.[31]   They are temporally progressive, each acting as the material determinant of the succeeding one. But Philip is not consistent as to whether a lower form itself behaves as matter toward a higher succeeding form, or whether it simply perfects matter to the point that it is properly adapted for the arrival of the higher form. They are one soul because

---

[27] "Et miro modo homo, cum habeat in se continuare superiora et inferiora, hanc habet in se triplicem secundum diversa; prima enim unitas est in ipsa anima secundum se considerata, secunda in corpore, tertia in toto homine." *Summa de bono* IV, Q. 8, ed. Wicki, II, p. 287; ed. Keeler, p. 88.

[28] V. Doucet, "A travers le manuscrit 434 de Douai," *Antonianum* 27 (1952), 531-80, on p. xxx.

[29] P. Glorieux, ed. *La 'Summa Duacensis' (Douai 434)* (Paris, 1955), p. 10 and "Les 572 Questions du manuscrit de Douai 434," *Recherches de théologie ancienne et médiévale* 10 (1938), 123-52, on pp. 112-34.

[30] Wicki, *ed. cit.*, Pars prior, p. 49*.

[31] *Summa de bono* IV, Q. 3, ed. Wicki, II, p. 233; ed. Keeler, p. 32.

the vegetative and sensitive are not in their subject *per se* but are only preparatory to the perfecting arrival of the rational soul, which alone is the soul in man.   Philip develops a theory of the progression of forms, some aspects of which are similar to, but also different from, that which Aquinas would later adopt: Man is not progressively a vegetable, an animal, and a rational being, because the two lower forms are not his perfection.   Each succeeding form adds a power to those of the preceding, until the perfecting form, the rational soul, arrives.   The resulting soul is a compound one, two thirds corruptible and one third incorruptible: "Just as sometimes a ray of fire and a ray of the sun are united together and are like one ray ... and just as it happens that the ray of fire is corrupted and that of the sun remains, thus it is in souls, since two are corruptible with the body, and the third remains and is separated from the others and from the body, as the perpetual from the corruptible, as Aristotle says."[32]   So the ray of fire (i.e., the vegetative and sensitive souls) continues to exist while the body is alive, but goes out when the body dies, and only the light of the sun (i.e., the rational soul) remains. But Philip never holds that the rational soul has other than rational powers. William of Auvergne will use the same example, but in his thought the lower forms are absorbed by the rational, which performs all the life functions, thus avoiding either a compound soul or a duplication of vegetative and sensitive souls.

Although Philip has read Averroes's Great Commentary on *De anima*, he does not seem to be aware of the Averroistic doctrine of the uniqueness of the possible and agent intellects, and he understands Aristotle as having taught that these are both aspects of the rational souls of each individual. and that they are separable in the sense of not operating by means of any bodily organ. This will be the common understanding of Averroes until the decade of the 1250s.[33]   Among the differentiae of intellect—possible, agent, and formal—

> only the formal intellect is destructible, because it was acquired and can be destroyed and is diverse according to diverse understood things.   But with respect to both differentiae, the other is incorruptible.   For the agent intellect is "quo omnia est facere" and the possible intellect

---

[32]   "Sicut quandoque radius ignis et solis simul uniuntur et non sunt nisi quasi unus ..., et sicut contingit quod radius ignis corrumpitur et remanet solaris, sic est in animabus, quoniam due corrumpuntur cum corpore et tertia manet et separatur ab aliis et a corpore, sicut 'perpetuum a corruptibili,' sicut vult Aristoteles." *Summa de bono* IV, Q. 3, ed. Wicki, II, p. 234; ed. Keeler, p. 33.

[33]   See Dominique Salman, O.P., "Note sur la première influence d'Averroës, *Revue néoscholastique de philosophie* 40 (1937), 205-12.

"quo omnia est fieri." And both intellects are separable. Indeed the agent intellect abstracts from matter and from natural dispositions, which it could not do unless it were separable. ... Likewise, concerning the possible intellect, it is shown that it is separable and incorruptible; as Aristotle says, it is unmixed and separable.[34]

Despite its shortcomings, Philip's doctrine of the soul was extremely influential, and echoes of it may be found in many succeeding writers. His formulation of a theory of a hierarchy of forms, each lower one acting as matter toward the succeeding form; his concept of the compound soul; his assertion that there is the least degree of unity in man; and his distinction among different kinds of form and different kinds of bodies are the most significant. Some or all of them can be found in the works of John of La Rochelle, Albert the Great, Anonymous Admontensis, Anonymous Vaticanus, Peter of Spain, John Pecham, Siger of Brabant, Matthew of Aquasparta, and Geoffrey of Aspall. Twelve manuscripts of the *Summa de bono* have survived from the thirteenth century alone. Philip must be considered a major, if not always salutary, influence on thirteenth-century discussions of the soul.

ALEXANDER OF HALES. Contemporary with Philip the Chancellor in the Parisian theological faculty was Alexander of Hales. He was highly thought of in his time, and he achieved even greater importance than he might otherwise have had by virtue of adopting the Franciscan habit in 1236, at the height of his career, thus establishing a Minorite presence in that faculty that would not be surrendered. After his death in 1246, his disciples put together a lengthy *Summa* purporting to represent his thought but unfortunately containing much inauthentic material. This *Summa* has been meticulously edited, but the task of sorting out Alexander's authentic work from that of others remains daunting. It is particularly desirable to know what Alexander himself taught, because later in the century, John Pecham would contrast the traditional Christian purity of the teaching of Alexander with the pernicious novelties that were corrupting theology in his own day. But what we can know with some confidence about Alexander's teaching indicates that he

---

[34] "... solus intellectus formalis est destructibilis, quia acquisitus est et tolli potest et secundum diversa intellecta diversus est. Sed secundum utramque differentiam alter incorruptibilis est. Intellectus enim agens est 'quo omnia est facere,' intellectus possibilis 'quo omnia est fieri,' et uterque <separabilis>. Intellectus quidem agens abstrahit a materia et materialibus dispositionibus, quod non posset nisi esset separabilis. ... De intellectu iterum possibili ostenditur quod separabilis et incorruptibilis, sicut dicit Aristoteles: 'Immixtus est, separabilis,' etc." *Summa de bono* IV, Q. 6, ed. Wicki, II, pp. 270-71; ed. Keeler, pp. 61-2.

himself was an adventurous teacher, eager to come to grips with the new material and not at all fearful of the newly translated works.[35]

Like most of his contemporaries, Alexander maintained that soul and body are two complete and distinct substances, and this standpoint led him to react to the Aristotelian material in interesting ways. He saw no incompatibility between the doctrines of Augustine and Aristotle, and he tried to harmonized them, usually by interpreting Aristotle in a sense consonant with Augustine. The relationship between body and soul, he says, is not completely similar to that between matter and form. "For the soul is a *hoc aliquid* apart from its matter, and one cannot say this of a natural form in an absolute sense. Whence [the soul] is not here, properly speaking, the act of matter, but the act of a natural body completed in its natural form, which form is called the corporeal form."[36] The union of soul and body is the union of one substance with another substance.[37] The soul is not the principle of the body insofar as it is a body, but of the body insofar as it is actually living.[38]

If soul and body are two substances, it is necessary to investigate the manner of their union. Alexander says that the rational soul is joined to the body as a mover to the thing moved and as the formal perfection to the thing perfected by it. In the first way, it has mediating powers by which it moves the body according as the body is an organ of the soul;[39] these powers are borne by vital and animal spirits, which then exercise them in various parts of the body. But insofar as the soul is the perfection of the body, it is united immediately, although in a sense the preceding dispositions that are responsible for the growth and organization of the embryo to the point that it is capable of receiving the rational soul may be called means. "But in the union itself, the soul itself is joined to the body."[40]

---

[35] See, for example, the way he dealt with the problem of the eternity of the world in R. C. Dales, *Medieval Discussions of the Eternity of the World*, pp. 65-70.

[36] "Est enim anima hoc aliquid praeter suam materiam: quod non est dicere in forma simpliciter naturali. Unde non est ibi proprie actus materiae, sed actus naturalis corporis completi in forma naturali, quae forma dicitur forma corporalis." Alexander of Hales, *Summa theologica* II, num. 347, ed. Quaracchi, II. p. 422.

[37] "Licet uniatur cum suis potentiis, per se tamen est unio substantiae ad substantiam corporis." *Op. cit.*, II, num. 112, *ed. cit.*, II, p. 151, ad 3.

[38] "[N]on est principium corporis quia corpus, sed corporis quia vivens in actu." *Op. cit.*, I, num. 46, *ed. cit.*, I, p. 73, ad 1.

[39] *Op. cit.*, II. num. 345, *ed. cit.*, II, p. 420.

[40] *Loc. cit.*

In investigating the question of whether there is life in the embryo before the infusion of the rational soul, Alexander tells us more about these powers. First he presents the alternative explanations of which he is aware: 1- some say that the embryo lives a vegetative and sensitive life by virtue of the vegetative and sensitive souls, before the infusion of the rational soul, and these people posit three souls in man; 2- others say that although there are three substances, these are nevertheless not three souls, because soul is the name of the perfection, which is the rational soul, while the other two act as material dispositions for it—but this position is contrary to Augustine, who says that there are not several souls in man; 3- others, supposing that the rational soul has its own vegetative and sensitive powers, posit that there are necessarily two vegetative and sensitive souls in the same being and that the prior ones cease to operate when the more powerful ones of the rational soul arrive—but this does not seem probable because nature does nothing in vain and because this position would also require more than one soul in man; 4- some people posit that the prior souls are corrupted when the rational soul arrives, and having performed their function, they cease to exist—and this does not seem probable either.

Alexander then presents his own view, which is based on Aristotle, *De generatione animalium* 2, 3, modified by Augustine, versions of which we shall meet again in the doctrine of Roland of Cremona and Peter of Spain. After the seed has been deposited by the father, Alexander says, a power remains in it, which becomes stronger and is invigorated by the natural powers of the womb that contains it. Thus strengthened, the power remaining in the semen prepares the body and renders it suitable for the reception of the rational soul, with the containing body of the mother all the time cooperating.

Therefore, he says, it is not necessary for there to be a vegetative and sensitive soul in the embryo prior to the rational in order to prepare it for receiving the rational soul, but only in the manner of an exterior faculty that prepares the body "which ought to be an organ of the perfecting soul itself or the thing that is perfected by the soul." Consequently, the embryo prior to the infusion of the rational soul does not have life properly speaking, but only through an inexact extension of the meaning of the word, as "apples are said to live, not because they are vivified by the presence of a vivifying and informing soul, but from the influence they receive from the plant to which they are attached."[41]

Like Philip the Chancellor, Alexander teaches that since the soul and body are so widely disparate in their natures, they can only be joined by a series of means, although he works out a somewhat different scheme from that of Philip. In Alexander's view, the extremes are the rational soul, a substance

---

[41] *Op. cit.*, II, num. 489, *ed. cit.*, II, pp. 582-83.

that is incorporeal, motive, cognitive, and separable or independent from the body; and the body to which it is united, which is composed of the four elements and possesses a definite complexion. Both the vegetative power, which is incorporeal, not cognoscitive, and inseparable, and the sensitive power, which is incorporeal, motive, cognoscitive, and inseparable, share incorporeity with the rational soul. Similarly, there are two things that share corporeity with the body: spirit, which is a body not composed of the four elements; and humor, which is a body composed of the elements but lacking a definite complexion. And consequently there are four means uniting the soul to the body, two from the side of the soul, and two from the side of the body.[42] Alexander does not, like Philip, refer to the vegetative and sensitive powers as substances, but he does consider that they nevertheless continue to exist and to function after the infusion of the rational soul. The sensitive power, he says, is a mean between the essence of the soul and the act that is accomplished in a different part of the body or in a different organ.[43]

This exaggerated dualism led him to understand Aristotle as positing two intellects in man, one immortal, the other perishable. "One is incorruptible, and this is the intellect pertaining to the soul itself; it is separable from the body, just as the soul is. There is another intellect, which receives the species abstracted from the phantasms. This intellect, so Aristotle says, is corruptible. The intellect of this kind is the act of the body and the principle of movement of the body."[44] This corruptible intellect, which Alexander calls the material intellect, is, according to him, "the act of man in the body," while the possible and agent intellects are powers of the incorruptible intellect.[45] He elsewhere locates this corruptible intellect in the "media cellula capitis,"[46] and seems to

---

[42] *Summa theologica* II, num. 346, *ed. cit.*, II, p. 421.

[43] *Op cit.*, num. 354, *ed. cit.*, II, p. 431.

[44] "Dicendum quod duplex est intellectus: unus qui est incorruptibilis, et iste est intellectus animae secundum se et separabilis a corpore sicut anima; alius est intellectus qui recipit species abstractas a phantasmatibus, qui est corruptibilis, ut dicit Philosophus, qui corrumpitur, quodam interius corrupto, et huiusmodi intelligere est actus corporis et principium movendi in corpore." Alexander of Hales, *Quodlibet III*, q. 6 (Vat. lat. MS 732, fol. 27d, printed in Roberto Zavalloni, *Richard de Mediavilla et la controverse sur la pluralité des formes. Textes inédits et étude critique.* Philosophes Médiévaux 2 (Louvain, 1951), p. 399.

[45] Alexander of Hales, *Summa theologica* II, n. 374, ed. Quaracchi, p. 454.

[46] *Op. cit.*, num. 348, *ed. cit.*, II, p. 423; see also num. 371, II, p. 450. The editors cite William of Conches, *De philosophia mundi*, IV, cap. 24 (*PL* 172:95). Another possible source is Costa ben Luca, *De differentia spiritus et animae*, cap. 2, *ed. cit.*, pp. 124-27.

imply that it coexists in man with the intellect that is part of the soul. It is this intellect that understands with phantasms; it differs from the similar faculty in beasts only because it is found in a subject that possesses intelligence properly so-called.[47]

It is difficult to know what to say concerning Alexander's views on a double vegetative-sensitive soul. In num. 682 he explicitly rejects it in the form in which he had stated it, but some such view seems to be implied in what he says elsewhere—for example that the vegetative and sensitive powers continue to function and act as means for the operation and union of the rational soul, and in his doctrine of the double intellect, which we have described above. Alexander says that although these powers are corruptible considered in themselves, since they are rooted in the substance of the rational soul and are dragged along with it after the death of the body, they are separable not *per se* but because of the substance in which they are rooted. But just how these powers came to be rooted in the rational soul, when they existed in the body before the infusion of the rational soul, is never made clear.

It is evident that Alexander was keenly aware of the difficulties involved in considering the powers responsible for the growth and organization of the embryo prior to the infusion of the rational soul to be souls, or substances in any sense. He tried to accept the main points of Aristotle's teaching without abandoning the Augustinian insistence on the unity of the human soul. He remains an extreme dualist in his view of the relationship of body and soul: his view is that the relationship is primarily an operational one, and even as he insists that, considered as the perfection of the body, the soul is united to it immediately, he admits media that make the union possible. He never calls the soul the form of the body, and while admitting some similarity between the soul-body relationship and the matter-form relationship, he emphasizes the dissimilarities.

WILLIAM OF AUVERGNE. William of Auvergne, bishop of Paris from 1228 to his death in 1249, wrote extensively on the soul, in his *De universo*,[48] composed for the most part between 1231 and 1236, and, in much more orderly fashion and at even greater length, in his treatise *De anima*,[49] which

---

[47] *Summa theologica* II, num. 332, *ed. cit.*, II, p. 404, ad 2. See Zavalloni, *op. cit.*, p. 400.

[48] In his *Opera omnia* I, pp. 593-1074 (Paris/Orleans, 1674; repr. Frankfurt a. M., 1963). There is no modern critical edition.

[49] *Opera omnia* II, suppl., pp. 65-228. Two important studies of William's *De anima* are E. A. Moody, "William of Auvergne and his Treatise *De anima*," *Studies on Medieval*

was probably his final composition. His efforts reveal a man who is earnestly and with great intellectual honesty trying to deal both with the traditional, essentially Augustinian, material and with the recently translated works of various Muslim, Jewish, and Greek authors, particularly Avicebron, Avicenna, Alexander of Aphrodisias, Aristotle, and Averroes. The result is often a quagmire of apparent contradictions, inexact analogies, unfinished arguments, and a capricious and inconsistent use of such technical terms from the Aristotelian vocabulary as form, matter, potency, substance, agent intellect, and so on. This makes it very difficult at times to divine what William is actually saying, and so to provide an accurate account of his doctrine. But there were two basic convictions that underlay all of his diffuse argumentation. First was the absolute simplicity of the soul, and second was the essentially active character of human cognition, although even these two positions seem sometimes to be contradicted by his words.

William's view of the body-soul relationship is, like that of his contemporary Alexander of Hales, an unabashed dualism; each is a distinct substance. The soul is not the substantial form of the body,[50] insofar as it is a body; that is, it is not the first act of the being of the body; this is a corporeal form. Soul and body are each complete substances, and the soul cannot be called, strictly speaking, the act of the body, but only in a manner of speaking.[51] The soul alone is the unique subject of all human operations; the body is simply its instrument. The relationship is like that of master and servant, or ruler and ruled, or operator and instrument.[52] The composite

---

Philosophy, Science, and Logic (Berkely/Los Angeles, 1975), pp. 1-109; and Steven Marrone, *William of Auvergne and Robert Grosseteste* (Princeton, 1983). His view of the body-soul relationship is intensively studied by Bernardo Bazán, "Pluralisme de formes ou dualisme de substances?", pp. 44-47, whose conclusions I have accepted in what follows.

[50]   See *De anima, ed. cit.,* pp. 196 and 199; Moody, *op. cit.,* pp. 43-44; and A. Masnovo, *Da Guglielmo d'Auvergne a San Tommaso* 3, p. 97.

[51]   "Cum enim dicat iuxta sermonem Aristotelis animam actum esse ejus potentiae, qua corpus dicitur potentiae vitam habens in ratione vel definitione animae ... actus autem ibi non intelligitur nisi perfectio." *De anima, ed. cit.,* p. 118a.

[52]   "Et quoniam longe nobilius impetrare quam servire, et imperativum omni ministrativo naturaliter praecellit; quoniam ad invicem se habent sicut dominus et servus, vel dominans et serviens, cum manifestum sit totum corpus humanum ad modum ministrantis se habere sive servientis ad huiusmodi operantem sive dominantem, imo quod minus est ut instrumentum ad operantem, et per illud manifestum est motorem huiusmodi hos motus imperantem longe nobiliorem esse ipso corpore humano, et propter hoc esse substantiam, cum accidens omne sua substantia longe ignobilius esse necesse est." *De anima, ed. cit.,* p. 68a. "Amplius manifestum est nullum instrumentum esse propter se, sed propter operatorem, ad hoc videlicet ut se serviat in operationibus quae fieri habent per ipsum. Cum igitur corpus humanum organicum sit, quod est dicere instrumentale, imo cum sit instrumentum unum ad

composed of body and soul does not involve the transcendental relation of matter and form but is purely instrumental, such as the relation between a knight and his horse.[53] As he understands Aristotle's definition of the soul as "the first perfection of a natural organic body, with the potentiality for life," this "natural organic body" must refer to the human embryo, since it could not possess life potentially if it already possessed it actually.

But at this point, William disagrees sharply with Alexander. While it is developing into a human body, the embryo has only its corporeal form, but when it has become a human body the rational soul is created in it, transcending and absorbing the previous form, as a greater light supersedes and absorbs lesser lights, which thereafter cease to exist, since they would be superfluous.[54] There is therefore only a single soul in each man. The soul is not composed of matter and form, not even spiritual matter and form, but is absolutely simple, although this one simple substance has a diversity of operations, all of which are properly those of the soul.

It would seem, from what we have seen so far, that humanity itself belongs only to the soul and that the body counts for nothing in a human being. William does not go quite so far. "Humanity," he says,

> is not the soul alone. But the soul is the perfection of the body—its essential perfection, I say. Therefore [the body] is part of [humanity], and with prime matter it composes and constitutes it. Thence it is clear that if the human soul were joined to a body made of air or iron so that from them

---

multas operationes aptum, natum et fabricatum, necesse est operatorem non esse cui naturaliter serviat, quique eo naturaliter uti debeat. Hic autem operator est quam vocamus animam humanam." *De anima, ed. cit.*, p. 67b.

[53] "Attende autem diligenter exempla quae tibi posita sunt de equo et equite, de domo et inhabitatore, de instrumento et operatore, de veste, seu vestimento atque vestito; et apparebit evidenter quod rationabilius est, ex sermonibus nostris, usuique loquendi consonantius ut corpus dicatur pars hominis quae cadit in rationem hominis in quantum hominis, quemadmodum equus in rationem equitis. Ipsae enim operationes quae fiunt per corpus, ut ostensum est tibi in praecedentibus, ipsius animae humanae verissime ac propriissime sunt sicut est loqui, disputare et etiam, quamquam indignetur Aristoteles, texere et aedificare." *De anima, ed. cit.*, pp. 101-02.

[54] "Si vero quaesierit quis, quid fiat de anima huiusmodi, cum anima rationalis embryoni advenerit, respondeo quod desinit esse: otiosum enim esset esse ipsius post adventum animae rationalis; causa autem in hoc est, quoniam anima rationalis sufficit regere ac vivificare plenissime corpus cui advenerit. Quod se quis dicere voluerit, quia sic fit de anima vegetabili per adventum animae rationalis, quemadmodum de luce minori cum longe maior supervenerit, non improbabiliter dixit; videtur enim velut absorberi lux minor a maiori vel extingui." *De anima, ed. cit.*, p. 107a.

there were one composite, it is clear that this [result] would
not be a human being, as undoubtedly would be the case if
the soul alone were humanity.[55]

William's argument against the pre-existence of souls well illustrates how
difficult it is to extract a coherent doctrine from his words:

Further, in the generation of bodies it is manifest that the
corporeal form is not generated outside of matter, but in
matter. Why therefore should it be different in the case of
man? That is, why should the form of the man, which is
assuredly the soul, come into being or be generated outside
of his matter, which is nothing else than the human body?
It is also obvious that generation is nothing other than the
generation of a form in matter, at least in those things
which are composed of matter and form [as the human
composite is]; hence the generation of a man is nothing else
than the generation of the incorporeal soul, since the soul
itself is the form of the body or matter.[56]

On the face of it, this sounds like an argument for the soul's being *ex traduce;*
although, because of his many contrary statements in other places, this cannot
be his meaning.   If we substitute for 'generation' the traditional terms
'creation and infusion,' William's account is not exceptional; but the
exigencies of his argument necessitated the use of the concept of generation as
well as the word, and this makes for misunderstanding.   And whereas he has
elsewhere emphatically denied that the soul is the form of the body, but rather
of the animated composite, he here flatly states the contrary.
    William wrestles mightily with the concept of the agent intellect, but again
his teaching is sufficiently ambiguous that Étienne Gilson writes that,
according to William, the soul's cognitive function cannot be the agent
intellect, which can only be God;[57] and Stephen Marrone states baldly that

---

[55] "Propter hoc humanitas non est anima sola, sed anima est perfecti corporis, perfectio inquam essentialis ipsius, quae est pars illius est, et cum materia prima componit et
constituit illud. Hoc inde manifestum est, quia si corpore aëreo vel ferreo conjungeretur
anima humana, ita ut ex illis esset unum compositum, manifestum est quod illud non esset
homo, quod indubitanter esset, si sola anima esset humanitas." *De anima, ed. cit.*, p. 66b.

[56] *De anima, ed. cit.*, p. 125, cited in Moody, *op. cit.*, p. 32.

[57] *History of Christian Philosophy in the Middle Ages* (New York, 1955), p. 256.

William "rejects the idea of an agent intellect in any form."[58] Texts could be found to support either of these interpretations.

William begins his investigation by asserting the absolute simplicity of the soul. Since it is simple, it is also indivisible, and so it is impossible that the agent intellect be a part of the soul. If the agent intellect exists, William argues, it must either be a spiritual substance distinct and separate from the soul, or it must be the soul itself. It cannot be the soul itself for two reasons: first, because if the soul were the agent intellect as well as the substance in which intelligible forms are actualized or received, this would destroy the distinction between potential and actual understanding that occasioned the positing of an agent intellect in the first place; and second, if the soul were the agent intellect, it would always be in act and would understand all intelligibles continuously, and this is clearly not the case. Neither can it be a separate substance distinct from the soul. William here attributes to Aristotle the Avicennan doctrine that the agent intellect is the tenth celestial Intelligence, the Giver of Forms, although he had clearly read Aristotle's *De anima.* In his argument against 'Aristotle,' William develops his own view. He criticizes the Aristotelian analogy of intellectual action and passion with corporeal action and passion. The human mind is not purely passive, receiving information from the senses, for this would reduce it to the status of a book that could not read itself or a mirror which could not see what was reflected in it. Cognition is essentially an active rather than passive process. The soul cannot be acted upon by the body, but it can attend to sensation and then produce intelligible forms from its own substance. It is essentially the same as the material intellect, since understanding takes place in it. It does not require an agent intellect to make actual understanding from potential understanding, for it generates these intelligible forms from within itself. This is possible because the soul was created with a knowledge of first principles innate to it. Hence, no other principle is necessary to account for the unfolding of knowledge in the soul by reasoning and philosophy. "Therefore, the positing of an agent intellect is completely futile and is only a figment."[59] The function of the Peripatetic agent intellect is performed by God in giving the first principles of truth to the soul, and it is manifested within the soul by virtue of these principles, not outside it. This explains why the intellect is not at all times completely in act.

The Augustinian element in William's thought is much more profound that it had been in that of the other authors we have considered to this point. In view

---

[58] *William of Auvergne and Robert Grosseteste*, p. 58.

[59] "[F]igmentum igitur est tantum, et vanissima positio intellectus agentis." *De anima, ed. cit.*, p. 209.

of this, the earnest effort he made to accommodate the new learning is all the more impressive; and although the result of this effort did little to solve the problems inherent in the basically incompatible views of the soul and of human intellection, it does serve to highlight them. It is of some interest that William thought very highly of Avicebron (whom he considered to be a Christian), Aristotle, and Averroes, all of whom he praises and whom he tries to understand, whenever possible, in a way consonant not only with Christianity, but with the Augustinian tradition. The net result of this, however, was a radical revision of that tradition. His work above all illustrates the disarray of scholastic thought on the soul during the 1230s and 1240s and confirms the picture we have derived from Philip the Chancellor and Alexander of Hales. William's position both as a member of the faculty of theology at Paris and as bishop of that diocese during a very difficult period of the university's history put him in touch with the major intellectual currents of his day. His writings have great value in reflecting these currents, although their influence on subsequent authors is not yet clear.

ROLAND OF CREMONA. Roland of Cremona, the first Dominican regent in theology at Paris (1229-30),[60] had less to say about the soul than William of Auvergne, and unlike William he maintained the matter-form composition of the soul.[61] However, he agreed with William in maintaining the position that there was only one soul in man and that it was simple, the vegetative, sensitive, and rational faculties all being performed by the single human soul. But he disagreed with William on the body's having its own form, which was absorbed by the rational, as well as with those who held that the embryo had some sort of life of its own before the infusion of the rational soul, which, following certain unnamed "physici," he placed on the forty-sixth day.[62] He asserted that the embryo received the principle of growth and development from its mother through the catyledons and did not have a life of its own until

---

[60]   On Roland see E. Prete, "La posizione di Rolando da Cremona nel pensiero medievale," *Rivista di Filosofia neoscolastica* 23 (1931), 484-89; C. R. Hess, O.P., "Roland of Cremona's Place in the Current of Thought," *Anglicum* 45 (1968), 429-77; and E. Filthaut, *Roland von Cremona O. P. und die Anfänge der scholastik im Predigerorden. Ein Beitrag zur Geistesgeschichte der älteren Dominikaner* (Vechta, 1936).

[61]   D. O. Lottin, "La composition hylémorphique des substances spirituelles," *Revue néoscholastique de philosophie* 34 (1932), 21-41, on p. 24.

[62]   "Dicunt enim physici, quibus est credendum in sua facultate, quod quadrigesimo et sexto die post conceptionem infunditur anima rationalis, quia tunc sunt formata omnia membra secundum quod competit animae rationali." *Summa theologiae*, Paris, Bibl. Mazarine MS 795, fol. 54ra, quoted by Hess, *art. cit.*, p. 439. This same work is designated as Roland's *Comm. in Sent.* by Zavalloni.

the infusion of the rational soul.[63] He also avoided using the word 'form' to designate the relation of body and soul, but called the soul instead the body's perfection, thus emphasizing the traditional distinction of body and soul as two distinct substances.[64]

Still, he emphasizes the soul's dependence on the body. It differs from an angel specifically by the fact that it is created with a dependence on the body, to which it naturally desires to be joined.[65] When the soul leaves the body, it is no longer a soul (although it is a spirit) because it loses the relationship to the body by which it is called a soul.[66]

ROBERT GROSSETESTE. It was not only at Paris that significant work on the soul was being done, for during the 1220s the theological faculty at Oxford was reaching a respectable level of achievement. Although several who taught there had also studied and taught at Paris, the intellectual life of the university during the 1220s and 1230s was dominated by a scholar peculiarly English in his education and his orientation.[67] Although Robert Grosseteste kept himself aware of what was going on at Paris, his own mind was highly original and idiosyncratic, so that his treatment of most questions was of a very different nature from that of mainstream theologians. Still, his influence

---

[63] "Sed frustra nituntur; quia embryo non crescit vel vegetat nisi vegetatione matris suae, quia antequam infundatur ei anima rationalis, est sicut quoddam membrum matris, quoniam continuatur matrici embryo per cotilidones. Sensibilis et vegetabilis sunt vires anime rationalis in homine... Dicunt quod in homine est anima vegetabilis et anima sensibilis et anima rationalis. Sed hoc non potest stare, quia unius rei unica est perfectio prima, quia unius rei unicum est esse... Si prima perficit, pro nichilo venit secunda vel tertia... Constat autem quod anima est perfectio huius corporis organici potentia vitam habentis. Ergo hec anima vegetabilis est perfectio huius corporis et hec anima sensibilis et hec anima rationalis." Roland of Cremona, *Commentarius in libros Sententiarum*, printed from Paris MS Mazarine 795, fol. 34v by Roberto Zavalloni O.F.M., *Richard de Mediavilla et la controverse sur la pluralité des formes* (Louvain, 1951), p. 387, n. 15. Cf. Aristotle, *De generatione animalium* 2, 1 (734b) and 2, 3 *passim* (= *De animalibus* 17).

[64] Roberto Zavalloni O.F.M., *op. cit.*, p. 387.

[65] "Angeli autem et omnes animae differunt secundum speciem ..., quia anima creatur cum dependentibus a corpore, et haec est substantialis differentia. Anima naturaliter appetit incorporari, id est, anima, quae est, habet in se dependentiam qua naturaliter vult incorporari." *Summa, MS cit.*, fol. 55vb, quoted by Hess, *art. cit.*, p. 438.

[66] *Ibid.*, fol. 87ra, quoted by Hess, *art. cit.*, p. 438.

[67] This is the provocative and convincing thesis of R. W. Southern, *Robert Grosseteste: The Growth of an English Mind in Medieval Europe* (Oxford, 1986).

on Oxford thought was powerful and lasting, and through the mendicant orders, whom he valued highly, on that of the continent as well.

Although a treatise *De anima* is attributed to Grosseteste by S.H. Thomson,[68] Fr. Leo Keeler has shown very clearly that this work is dependent on the teaching of Philip the Chancellor,[69] and he considers it a *reportatio* of Philip's lectures. But Keeler accepted the attribution to Grosseteste. D.A. Callus subsequently showed that the treatise is in fact derived from the *Summa de bono*, and so the problem of dates is virtually unsurmountable.[70] There are in fact many reasons of date, doctrine, and nature, which make it all but impossible that Grosseteste was the author of this *De anima*. I consider the treatise inauthentic and shall not include it in the discussion of Grosseteste's teaching on the soul.

Although Grosseteste was unusually well-read, he was never bullied by his sources. He derived from them information and suggestions, but he never hesitated to alter these in highly original ways. Even Augustine was used in this fashion, so it is not surprising that he felt no duty to be faithful to the exact meaning and terminology of Aristotle, even as he borrowed much from him.

The central feature of Grosseteste's view of the human soul is that it is the image of the Trinity; consequently its relation to the body imitates on a finite scale the relationship of God to his creation. As God is wholly present without site or place everywhere in the universe, so the soul is present wholly, without site or place, everywhere in the body. And as God is united to his creation through the unity of the person of Christ in the Incarnation, so the human body and soul are united in the unity of person in the individual human being. The soul is a substantial form, wholly spiritual and completely without matter. It is simple, and all the processes of human life result from its care for the body that it rules. It is created from nothing by God (as are the vegetative and sensitive souls of lower forms of life) and is infused into the body at conception. The soul operates on the body by means of light, the corporeal substance closest to the spiritual. Light in turn affects the nerves, muscles, and so on, of the body.[71]

---

[68]   "The *De anima* of Robert Grosseteste," *The New Scholasticism* 7 (1933), 201-21 and *The Writings of Robert Grosseteste* (Cambridge, 1940), p. 89.

[69]   Leo W. Keeler, S. J., "The Dependence of Robert Grosseteste's *De anima* on the *Summa* of Philip the Chancellor," *The New Scholasticism* 11 (1937), 197-219.

[70]   "Philip the Chancellor and the *De anima* ascribed to Robert Grosseteste," *Medieval and Renaissance Studies* 1 (1941), 105-27.

[71]   In his earlier works, Grosseteste expressed the commonly held view that the rational soul was infused at some point after conception, but his mature opinion, as expressed in the

Some aspects of Grosseteste's doctrine of the soul and the nature of human life have been quite well studied. His theory of human nature has received a magisterial treatment by James McEvoy.[72] His teaching on the natural powers of the soul, investigated by Servus Gieben,[73] holds that the soul has the power naturally to know all things and to direct its actions properly so that it might even attain a vision of God, although in most men this capacity is lulled or impeded by the corruption of the flesh or by a disorderly passion; and in any case the natural powers by themselves can go astray, fall into error, and do what is contrary to their nature, so that they must be perfected by grace. Still, this is a very high evaluation of the natural capacity of the human soul.

This is not to be wondered at, since over and over in his writings Grosseteste tells us that man is the highest of God's creatures, higher even than the angels. It is fitting, he says, that man should have been assigned a role of central importance in God's plan of reconciliation with his whole creation. Consummating a long but relatively submerged theme in Latin thought, extending from Gregory the Great through Eriugena to a number of twelfth-century writers,[74] Grosseteste claimed that the Incarnation would have taken place even if Adam had not sinned, because man, in the unity of his soul and body, summed up the entire created universe from the four elements and their qualities to the angels, so that God was able to reconcile himself with all of creation by becoming man, who was, in a sense, *every creature*. The rational soul and an angel, he says,

> share in the nature of rationality and understanding, and by
> this sharing they have an unbreakable chain and bond of

Hexaemeron and the sermon *Ex rerum initiatarum*, was that it is infused at conception. For a discussion of this point and for a fuller and more nuanced treatment of Grosseteste's doctrine of the soul, see James McEvoy, *The Philosophy of Robert Grosseteste* (Oxford, 1982), especially chapters 3 and 4.

[72] "Robert Grosseteste's Theory of Human Nature. With the Text of His Conference, *Ecclesia Sancta Celebrat*," *Recherches de théologie ancienne et médiévale* 47 (1980), 131-87. See also James McEvoy, "Grosseteste on the Soul's Care for the Body: A New Text and New Sources for the Idea," *Aspectus et Affectus. Essays and Editions in Honor of Richard C. Dales*, ed. Gunar Freibergs (New York, 1993), 37-56.

[73] "Le potenze naturali dell'anima umana secondo alcuni testi inediti di Roberto Grossatesta," *L'Homme et son destin*. Actes de premier congrès international de philosophie médiévale (Louvain/Paris, 1960), pp. 437-43.

[74] See Richard C. Dales, "A Medieval View of Human Dignity," *Journal of the History of Ideas* 38 (1977), 557-72.

natural unity. But the rational soul cannot share with the body in species and nature. Nevertheless, the rational soul is naturally suited to be the perfection of the organic body and is united to this body in the unity of person. Hence an angel and the rational soul are bound by a natural unity; but the rational soul and the human body are joined in a personal unity. The human body shares its nature with all natural bodies, because heavenly bodies share with the element fire the nature of light; fire and air the nature of heat; air and water the nature of humidity; water and earth the nature of cold. And the human body consists of the four elements, and therefore it shares the natures of these things, and as a consequence with the heavenly bodies in which fire shares the nature of light. Consequently it also shares with all elemented natures [i.e., natures composed of the four elements] sharing with the natures themselves. The rational soul also shares with the sensitive soul of beasts the sensitive power, and with the vegetative soul of plants the vegetative power. Therefore, man also shares the nature of all creatures, but the totality of creatures does not yet have unity with the creator, nor can the creator have unity of genus, species, or nature with the creature. But he can have unity with the creature so that he may assume the creature in the unity of person. Therefore, if God should assume man in the unity of person, the universe is brought back to the completeness of unity. But if he were not to assume human nature, the universe has not been brought back to the completeness of unity that is possible for it. Quite apart from saving fallen man, it was suitable for God to assume the unity of person, since God could also do that. Nor was it unfitting for him to do this, but it was utterly suitable, since without this the universe would lack unity. But when this was done, the universe would have the fullest and most seemly unity. And by this, all creatures would be brought back in a full circle, because without God's assumption of man in the unity of person, we are able to discover the bond uniting everything from angel to man, but not yet the bond of God to man, the last created thing, or to the angel, the first created. But when man has been assumed in the unity of person, now the bond is a complete circle.[75]

---

[75] "Anima autem rationalis et angelus communicant in natura rationalitatis et intelligencie, qua communicacione habent indissolubile vinculum et concatenacionem naturalis unitatis. Anima autem rationalis non potest communicare cum corpore in specie et natura. Est tamen anima rationalis apta nata ut sit perfeccio corporis organici, et uniantur ei in unitatem persone; quapropter angelus et anima rationalis concatenata sunt per unitatem

When he assumes human nature in the unity of person,
then the circle of creatures to the creator is most tightly
joined. And man having been assumed in the unity of
person, God is inserted into this same circle and is made the
beauty and honor of this circle like a gem in a ring of gold
... Nevertheless, the assumption of man by God the Word
unites the entire universe of things, because man shares
with animals sensation and with plants vegetation, and he
has a natural community with all other bodies through his
assumption in personal unity; and thus all things have a
fuller coupling with the creator in the assumed man. In this
coupling was made not only one universe of creatures, but
one universe of all things.[76]

---

naturalem. Anima vero rationalis et corpus humanum conveniunt in unitatem personalem.
Corpus autem humanum habet communicationem in natura cum omnibus naturis
corporalibus, quia corpora celestia communicant cum igne elemento in natura lucis; ignis et
aer in natura caloris; aer et aqua in natura humiditatis; aqua et terra in natura frigiditatis.
Corpus autem humanum constat ex quatuor elementis, quapropter communicat in natura cum
illis, et per consequens cum celestibus corporibus cum quibus communicat ignis in natura
lucis. Communicat etiam per consequens cum omnibus naturis elementatis communicantibus
cum ipsis elementis. Communicat quoque anima rationalis cum anima sensibili brutorum in
potentia sensitiva, et cum anima vegetabili plantarum in potentia vegetativa. Quapropter et
homo communicat in natura cum omni creatura.

Sed nullam habet adhuc universitas creature cum creatore unitatem, nec habere potest creator
cum creatura, ut dictum est, unitatem generis vel speciei sive nature; sed potest tamen habere
hanc cum creatura unitatem ut assumat creaturam in unitatem persone. Si igitur assumat Deus
hominem in unitatem persone, reducta est universitas ad unitatis complementum. Si vero non
assumat, nec universitas ad unitatis complementum sibi possibile deducta est. Circumscripto
igitur hominis lapsu, nichilominus convenit Deum assumere hominem in unitatem persone,
cum et hoc possit facere nec dedeceat ipsum hoc facere; sed multo magis deceat, cum sine
hoc careat universitas unitate. Hoc vero facto, habeat universitas plenissimam et
decentissimam unitatem, redacteque sint per hac omnes nature in complementum circulare;
quia sine eo quod Deus assumat hominem in unitatem persone, est reperire modo supradicto
concatenacionem quandam ab angelo usque ad hominem. Sed nondum est concatenacio Dei
ad hominem, ultimo creatum, vel ad angelum, primo creatum. Assumpto vero homine in
unitatem persone, iam est completa circularis concatenacio." *De cessatione legalium* III, i,
27-28, Richard C. Dales and Edward B. King, ed., *Robert Grosseteste De Cessatione
Legalium.* Auctores Britannici Medii Aevi 7 (London, 1986), pp. 130-131. The same is
contained more briefly in the *Hexaemeron* IX, viii, 2-3, Richard C. Dales and Servus Gieben,
O.F.M.Cap., ed., *Robert Grosseteste Hexaëmeron.* Auctores Britannici Medii Aevi 6
(London, 1982), p. 276.

[76] "[S]ed cum assumit humanam naturam in unitatem personae, tunc est circulus
creaturarum firmissime Creatori coniunctus, cum ipse Creator per unitatem personalem
assumpto homine in unitatem personae, sit eidem circulo insertus, factusque decor et honor
huius circuli tamquam aurei annuli ... [N]ichilominus tamen hominis a Verbo Dei assumptio
maxime rerum unit universitatem, quia, ut dictum est, homo et animalia in sensu, homo cum

An important aspect of this essay is that Grosseteste finds the unifying factor of body and soul in 'person.' The human personality then lies not exclusively in the soul, but in the unity of body and soul, which constitutes one person.

There is another aspect of Grosseteste's doctrine of the soul that has attracted very little attention. He had a much stronger feeling for the unity of body and soul than we usually encounter in medieval thinkers, and, as James McEvoy has pointed out,[77] he concentrated his attention on the concept of life, including soul and body, rather than on the soul alone as a separable substance. The soul's overriding natural desire is to be united to its body, to nurture it and care for it, to give it life. This view is expressed in several of his writings, but particularly in his gloss on 1 Corinthians 15:55 and in his *De cessatione legalium.*

In the first of these works, commenting on the text *Ubi est mors victoria tua?* Grosseteste composed a brief essay in which he emphasized the soul's desire to preserve and repair the body and to be joined to it. This was prevented only by the body's inexorable flow toward dissolution, which in this life the soul is powerless to prevent. And so ultimately the soul, against its will, is separated from its own body, and this is the victory of death. After the death of the body, the soul does not lose its desire to be joined to it and minister to it. But now the soul's desire is for an "embrace of perfection," when it "shall embrace the stability of the perfect body," which lacks any tendency toward dissolution. In this embrace death will not be joined, and "then there will be life in victory." The soul now sustains the body in stable perfection, in which, "as it were, privation is absorbed by possession, emptiness by fullness, defect by perfection." The body's death was the

---

plantis in vegetatione, cum aliis corporibus omnibus habens naturalem communicationem <per> assumptionem in unitatem personalem omnia, in assumpto homine pleniorem habent cum Creatore copulationem in qua copulatione facta est non solum una universitas creaturarum, sed et una universitas rerum omnium." Sermo *Exiit edictum,* edited from Brit. Libr. MS Royal VII F 2, folios 76va-77vb by Dominic J. Unger, O.F.M.Cap., "Robert Grosseteste Bishop of Lincoln (1235-1253) On the Reasons for the Incarnation," *Franciscan Studies* 16 (1956), 1-36, on pp. 22-23. Grosseteste's source for the doctrine that body and soul are unified in the unity of person is Augustine, *Epist. 137 Ad Volusianum* 3, 9-11 (*CSEL* 44, p. 110). It is also cited by William of Baglione, "Si illud quod constituit hominem in esse specifico operatione naturae de potentia materiae vel per infusionem animae rationalis," ed. Ignatius Brady, "Background to the Condemnation of 1270: Master William of Baglione, O.F.M.," *Franciscan Studies* 30 (1970), 5-48, on p. 23.

[77] James M. McEvoy, "Robert Grosseteste's Theory of Human Nature."

greatest punishment that could be suffered by the soul, since it entailed the dissolution of its union with the body, which is the soul's strongest desire.[78]

In two other works, Grosseteste develops this line of thought further. In the magnificent sermon *Ex rerum initiatarum*, he states that the soul's desire for union with its still-sound body, while the blood and vital heat are still abundant, is so strong that no created power is capable of effecting its dissolution.[79]   And in *De cessatione legalium* III, vi, 8-9 he puts forth, as clinching proof that the crucified Jesus was indeed God, an expanded role for the soul's care for the body. He claims that Christ died not as a result of the wounds inflicted during the crucifixion, but that he rather voluntarily sundered his soul from his body. And, he asserted, it is beyond any created power to separate the human soul from its still-whole and healthy body, since

> the soul naturally desires to be joined to its own body and abhors nothing more than its separation from the body through death. Whence it itself is naturally inseparable [from the body], provided that the vital heat is not yet deficient in the heart. And so it is not within human power to separate the soul from its body so long as it is still healthy and still possesses the vital heat; indeed, it requires no less power to separate it from a still-healthy body than to couple it to an organic body. Therefore, it is a proper work of divine force and creative power voluntarily to separate a soul from its still-healthy body. And so when the Lord Jesus, his body still healthy, hanging on the cross, voluntarily sent forth his own spirit, he performed a work that was divine and proper to divinity alone, since he died with his interior and vital members still sound, and with the vital heat remaining.
>
> It is evident that he was crucified while he was still healthy and whole, fastened to the cross with the nails perforating only his hands and feet, his body not yet in any way wounded nor blood yet flowing out through any wound, because he died after spending only three hours on the cross. For the perforation alone of his hands and feet would not empty the blood of his heart and interior in so short a time, nor would it have extinguished the vital heat

---

[78]   Edited by Richard C. Dales, *Robert Grosseteste's Glosses on the Pauline Epistles*. Corpus Christianorum, series latina, continuatio medievalis (Brepols, Belgium, in press).

[79]   Servus Gieben, O.F.M.Cap., "Robert Grosseteste on Preaching. With an edition of the Sermon 'Ex rerum initiatarum' on Redemption," *Collectanea Franciscana* 37 (1967), 100-141, on p. 132.

of a strong and healthy youth so quickly. Another thing
that attests to the fact that he was not dying because of a
great flow of blood through the wounds made by the nails
is that when his side was pierced by a lance after he was
dead, blood poured out from his insides. A dead body that
has suffered no effusion of blood, if it were wounded after
death, would not usually bleed freely, for the blood would
be cold and coagulated. Furthermore, he died shouting in a
loud voice, not sighing but shouting forth, literately and
significatively, and suppliantly praying to God the Father,
saying: *Into your hands I commend my spirit*. But if he had
been only a man, and the blood and vital heat had been
spent, he would in no way have been able to shout like this.
Therefore, either the vital heat had not yet been exhausted,
or he himself was more than a pure man. If he was more
than a pure man, I win the point—that he himself was
God—because nothing is naturally superior to man except
God. But if the vital heat had not yet been exhausted in
him, he did not die because of the violence of the wound,
but by the power which can couple the soul to the body and
separate the soul from a body not yet depleted of nature and
vital heat. This power is not human, but only, as has been
said, divine. Whence also the centurion, seeing that he
expired shouting thus, said: *Truly, this man is the son of
God*.[80]

---

[80] "... cum anima naturaliter appetat coniungi suo corpori, nichilque tamen abhorreat
quam a corpore suo per mortem separacionem. Unde et ipsa naturaliter inseparabilis est, dum
in corde non dum defecerit calor vitalis. In humana itaque potestate non est animam suam a
corpore suo, adhuc sano et vitalem calorem adhuc habents, deponere. Immo minoris
potestatis non est eam a corpore sano manente adhuc same deponere quam eam corpori
organico cupulare. Divine igitur virtutis et potentie creatricis opus proprium est animam
suam a corpore suo adhuc manente sano propria voluntate deponere. Dominus itaque Ihesus
cum adhuc corpore sano in cruce pendens, voluntarie proprium emisit spiritum, opus fecit
divinum et divinitati soli proprium. Quod autem adhuc sanis interioribus et vitalibus membris
et calore naturali manente moriebatur. Inde manifestum esse potest quod cum sanus et integer
cricifixus esset, in ligno crucis affixus, clavis solummodo manus et pedes eius perforantibus,
et corpore suo nondum aliter vulnerato, nec sanguine per aliud vulnus ab interioribus
defluente, post diei horas tantum in crucifixione peractas obiit.

Non enim perforacio sola manuum et pedum in tam brevi temporis spacio sanguinem cordis
et interiorem evacuaret, nec vitalem calorem hominis fortis iuvenis et sani tam cito
extingueret. Attestatur quoque huic quod non moriebatur, scilicet propter multam sanguinis
per vulnera clavorum effusionem quod, aperto eius latere cum lancea postquam fuit mortuus,
exivit sanguis de interioribus, cum corpora mortua etiam sine sanguinis effusione, si
vulnerentur post mortem, non consueverint sanguine fluere, infrigidato et coagulato sanguine.
Preterea, voce magna clamans, expiravit; nec clamans clamore gemitus, sed clamose
prolacione vocis, litterate et significative et Patrem Deum suppliciter deprecantis, dicens: *In*

Other writers emphasized that the soul had a natural inclination toward the body, and some even held that the soul was incomplete while it was separated from the body. But no one stated so strongly the soul's natural desire for union with its body. It was this emphasis, and the attribution of the union of body and soul to the unity of person, in which Grosseteste's thought was most original and most influential. It was however too far removed from the scholastic mainstream to exercize a preponderant influence during the thirteenth century.

By the end of the 1230s, the scholastics had concluded their initial confrontation with the flood of exotic works, both Semitic and Greek, which would fructify and confuse European thought for the next century. They reacted in various ways. Blund admitted that the soul was the form of the body, but anxious to preserve the immortality of the soul, he accorded it only an accidental relationship to the body. He denied the matter-form composition of the soul, but instead of appealing to Boethius's *quod est* and *quo est* distinction, he found its composition to be of genus and differentia. Nequam denied outright that the soul was a form; it was a substance, and he realized that it could not be both. William of Auvergne, Alexander of Hales, and Philip the Chancellor all denied that the rational soul was the form of the body, a status they assigned to the sensitive soul or to the corporeal form. Roland of Cremona attempted to avoid the implications of Aristotelianism by borrowing a suggestion from Aristotle and asserting that the embryo had no sub-rational soul of its own but grew and developed by virtue of the mother's vital powers. Robert Grosseteste rode roughshod over the 'New Aristotle,' even though he was a leader in its dissemination, and while asserting the traditional doctrine that the soul was a spiritual substance, emphasized its natural affinity for the body; and ignoring the problems that a close investigation of Aristotle's thought would have involved him in, found a solution to the problem of human unity in an analogy with the Incarnation, which flowed from his concentration on the human soul as the image of the Trinity.

---

manus tuas commendo spiritum meum. Si autem purus homo fuisset et defecissent in interioribus sanguis et calor vitalis non fuerat exinanitus, aut ipse fuit plusquam homo purus, aut simul sine caloris vitalis defeccione plusquam homo purus. Si fuit plusquam homo purus, tunc habemus propositum quod ipse fuit Deus, quia nichil superius homine naturaliter nisi Deus. Si autem calor naturalis nondum fuerat in illo exhaustus, non obiit per violentiam vulneris sed potestate que potest animam corpori copulare et animam a corpore nondum defecto a natura et calore vitali deponere. Que potestas humana non est, sed solum, ut dictum est, divina. Unde et centurio, videns quia sic clamans expirasset, ait: *Vere hic homo Filius Dei erat.*" *De cessatione legalium* III, vi, 8-9, *ed. cit.*, pp. 150-51.

By 1240 the tradition was in disarray. There was no consensus on what the human soul was or how it was related to the body, and the persistence of dualistic thought frustrated efforts at achieving a satisfactory account of human unity. For the next two decades, artists and theologians would grapple more or less successfully with the problem of trying to harmonize two discordant doctrines of the human soul.

CHAPTER THREE

TWENTY YEARS OF CONFUSION: OXFORD

Although, during the generation following Grosseteste's retirement from the schools to become bishop of Lincoln in 1235, no master approaching him in intellect or influence graced the Oxford faculties, nevertheless a group of masters of significant ability came to maturity between about 1240 and the mid-1250s. None of these has received the amount of scholarly attention he deserves, and most of that has been accorded by other than English scholars. There were, of course, many Englishmen, such as Alexander of Hales, Roger Bacon, and John Pecham, whose careers were primarily associated with Paris, even though they may also have taught at Oxford. And of the four masters we shall consider in this chapter, at least one, Richard Rufus of Cornwall, made quite a name for himself at Paris as well as Oxford. But the works we shall be considering of these men came out of the Oxford milieu and have much in common. Perhaps the most striking common trait is their clear perception of the incompatibility between what they considered to be Augustine's teaching (although most of the works they cite for this purpose are not by Augustine; for example, Gennadius of Marseilles, *De ecclesiasticis dogmatibus*, which they cite as Augustine, *De diffinicionibus recte fidei*, and Alcher of Clairvaux's *De spiritu et anima*) that the human soul is one substance with a variety of powers, and their understanding of Aristotle's teaching, that the vegetative and sensitive souls are educed from the potency of matter and are corruptible, and are distinct from the rational soul, which comes from without and is incorruptible. They exhibit much concern about the problem and are uniformly reluctant to take a clear definitive position on the question. They are all influenced to some extent (Adam of Buckfield less than the others) by universal hylomorphism and the plurality of forms, but none of them seems to be aware of a problem on the subject. They use Averroes extensively and seem to be unaware of his teaching on the possible intellect.

ADAM OF BUCKFIELD. During the early 1240s, two Oxford masters in particular are worthy of note, one an artist, Adam of Buckfield, the other a theologian, Richard Fishacre. If it is indeed true that the introduction of the "New Aristotle" proceeded apace at Oxford, while it was hampered by ecclesiastical prohibitions at Paris, much of the credit must go to master

Adam of Buckfield.[1]   Although it is possible, and even likely, that Adam eventually took a degree in theology, we have no evidence of this, and his Aristotelian commentaries, with one of which we are here concerned, surely date from his regency in arts, which had already begun by 1243.   Father Daniel Callus had begun an edition of Adam's commentary on *De anima* on the basis of the nine known manuscripts, and he printed sections of the commentary in an article in *Revue néoscolastique de philosophie*,[2] but he had not completed this at the time of his death, and so far as I know, no one has resumed his task.

Adam's purpose, as a professor of the arts, was to explain Aristotle's meaning, and for this purpose he depended heavily on Averroes.   As Fr. Callus has noted: "Comment after comment is often a mere paraphrase of Averroes's interpretation."[3]   It is clear that he had read the Commentator carefully.   It is somewhat surprising therefore that, like his contemporaries, he misunderstood him on a crucial point.   Adam does not even consider the question of a separate possible intellect, which would be so important after 1266.   Rather he confines his investigation to whether the agent intellect is a separate substance or a power of the soul.   He attributes the latter view to Averroes and the former to "many theologians."

There is a doubt, he says, whether the agent and possible intellect are of the same substance, or whether the substance of the possible intellect is within the soul, while that of the agent intellect is outside it.   Some people say it is the first way, that the substance of both is one and the same.   This substance is called the possible intellect insofar as it has an inclination toward the imagination, from which it receives intelligible forms, for it is thus bound to matter in some fashion.   But they call the same substance, considered in itself, the substance of the agent intellect.   This is evident from the text of *De anima*: "Necesse est in anima has esse differentias."   And the same point can be confirmed by many arguments, both of Aristotle and Averroes.   But many theologians are of the opinion that the agent intellect in us is the intellect of the First, whose intellect is indeed the true light, and that this light is more intimate to our soul than the soul is to itself.   They also base this opinion on Aristotle's statement that the agent intellect is a condition like light, and on

---

[1]   Martin Grabmann, "Die Aristoteles Kommentatoren Adam von Bocfeld und Adam von Bouchermefort.   Die Anfänge der Erklärung des "Neuen Aristoteles" in England," *Mittelalterliches Geistesleben* (2 vols., Munich, 1936) II, 138-87.

[2]   Daniel A. Callus, O.P., "Two Early Oxford Masters on the Problem of Plurality of Forms.   Adam of Buckfield—Richard Rufus of Cornwall," *Revue néoscholastique de philosophie* 42 (1939), 411-45, on pp. 433-38.

[3]   Callus, *art. cit.*, p. 419.

Averroes's comment that the substance of the agent intellect is its own action, and this would be true only of God.[4]

Adam shared with his Oxford colleagues a concern for and hesitancy about the unity of the soul, and he presents arguments for both sides. In his view, Aristotle had not settled the question; his remarks near the end of book 1 (I, 5, 411b 6-14, text 7 of the medieval version) were simply designed to refute Plato's contention that the various functions of the soul, such as understanding, opinion, desire, et cetera, pertained to different parts of the soul located in different organs, claiming instead that they all pertained to the soul as a whole, and that life resides in the soul as a whole.

But some people, he says, think that Aristotle was here treating the difficult question of whether the vegetative, sensitive, and rational souls are distinct only with respect to their operations, or also with respect to a diversity of substance. They are mistaken in this, Adam holds, and the latter question is still open for discussion.

He then gives a summary of the current views on the subject, beginning with five arguments why there must be only one substance for the soul. He first points out that one single being can have only one perfection, and since the soul is man's perfection it must be one in substance. The second argument is interesting in that it seems to imply that the body itself is not a substance, although Adam does not go into this matter: "One thing in act cannot be made of two things already in act. How then could there be more than one soul in man, or more than one substance in the soul?"[5] The third argument is that if the vegetative, sensitive, and intellective were of diverse substances, it does not seem possible to say how they could exist at once in one soul. Fourth, he asks what holds these three together if they were diverse substances. Not the body—rather the contrary. If the soul, then one could ask again whether the soul is undivided or not; and so we would have an infinite regress. The fifth argument is indebted to John Blund and Philip the Chancellor: It appears that the intellectual soul in man is the perfection of the sensitive soul, and the sensitive of the vegetative, in the same manner as the form 'differentia' perfects and completes the form 'genus.' But since genus and differentia do not differ except as complete and incomplete, it seems that the vegetative, sensitive, and intellective souls differ only according to completion and incompletion.

---

[4] Dominique Salman, O.P., "Note sur la première influence d'Averroès," *Revue néoscholastique de philosophie* 40 (1937), 203-12, on pp. 211-12.

[5] I have borrowed Fr. Callus's summary of Adam's Latin from "Two Early Oxford Masters," p. 437.

This is followed by four arguments for the opposing view. The first says that Aristotle considers all the powers of the soul except for the intellectual to be corruptible both in substance and operation and to be educed from the potency of matter, while the intellective is incorruptible with respect to it substance, even though not to its operations, and it comes entirely from without. Therefore, since the same substance cannot be at once corruptible and incorruptible, the substance of the intellective soul must be other than that of the vegetative and sensitive. The second argument notes that in *De animalibus (De generatione animalium* 2, 3 [736a] = *De animalibus* 17) Aristotle said that there was a power in the semen by which the members of the body are formed, and this potency afterwards becomes the act or form that is the vegetative soul. And this soul precedes the infusion of the intellective soul both by nature and by time. The third argument extends this line of reasoning to the sensitive soul; and since they both precede the intellective soul by time as well as nature, it is clear that the substance of the intellective soul is other than that of the vegetative and sensitive. And the fourth argues that since nature cannot produce anything higher than the vegetative and sensitive souls (since they can be educed from the potency of matter, while the intellective cannot), the intellective must come from something other than nature. And if this is so, the creator would not create it only according to its operations, but also according to its substance, which must therefore be other than the substance of the others.

At this point Adam interjects:

> And it can be said about this [i.e., that they are not of the same substance] that it is true and that the arguments for it are convincing, as some people hold, although Augustine and several others seem to be of the opinion that they are not diverse substances. For some hold that there is one and the same substance of the intellective, sensitive, and vegetative in man; the same, I say, in root and in essence, differing only according to operation,[6]

just as heat sometimes liquefies and sometimes congeals, depending on the nature of the matter. But Adam is suspicious of this position and does not see how those who hold it can reply to the arguments for the contrary position that have just been given, or perhaps some even stronger ones that might be

---

[6] "Et potest dici ad hoc, quod hoc verum est et quod raciones ad hoc procedunt, sicut volunt quidam: quamvis velit Augustinus, ut videtur, et plures alii, quod non sint substancie diverse. Volunt enim quidam quod eadem est substancia intellective, vegetative, sensitive in homine: eadem dico in radice et in essencia, differens solum secundum esse et inclinacionem ad opus." Edited by Callus, "Two Early Oxford Masters," p. 437.

devised. Some people attempt to do so, he says, by arguing that although the substance of the intellective soul is different from that of the vegetative and sensitive, "nevertheless these diverse substances are one soul and one perfection, as is illustrated by the fact that in a ray of the sun or of fire there are light and heat, which are diverse according to substance; and nevertheless it is one ray."[7] This is similar to the example we have met in Philip the Chancellor and William of Auvergne, but not quite the same. There is no implication in this example that the weaker ray is absorbed by the stronger, but rather that two distinct substances may be included in one substance.[8]

As J. M. Da Cruz Pontes notes, Adam hesitates to take a position,[9] but even though hesitantly he indicates his own preference by replying to the arguments in favor of one substance of the soul. And this preference is confirmed by his comment on book 2, 1, (413a 4-10), where Aristotle says that some parts of the soul cannot be separated from the body, but others can. Aristotle clearly means, he says, that "the intellective soul is separated from the body according to its substance, but the vegetative and sensitive are not. By this it is clear that it is not the same substance, but one and another."[10]

Among his replies to the arguments for the side he rejects, that to the fourth—that the vegetative, sensitive, and intellective would have to be unified by something if they were more than one substance—Adam reveals his debt to Philip the Chancellor:

> To the fourth it can be said that they are united by
> themselves and do not require another unifying agent, just
> as is the case with form and matter; for the intellective soul
> is in some manner the act and perfection of the sensitive,
> and the sensitive of the vegetative. For that which is prior
> among souls is in potency with respect to that which is
> posterior and is in some manner perfectible by it. And that
> the sensitive power is in potency with respect to the

---

[7] "Iste tamen substancie diverse sunt una anima et una perfeccio, sicut patet quod in radio solis et ignis sit splendor et calor, que diverse sunt secundum substanciam, et tamen est radius unus." *Loc. cit.*

[8] On this see J. M. Da Cruz Pontes, "Le problème de l'origine de l'âme de la patristique à la solution thomistique," *Recherches de théologie ancienne et médiévale* 31 (1964), 175-229, on p. 397.

[9] Da Cruz Pontes, *art. cit.*, p. 208.

[10] "Relinquitur igitur quod vult animam intellectivam secundum suam substanciam separari a corpore, substanciam autem sensitive et vegetative non; per quod patet quod non est eadem substancia, sed alia et alia." Callus, "Two Early Oxford Masters," p. 438.

intellective is clear from the fact that the imagination, which is the highest sensitive power, is more perfect and more determined in man than in other animals; and this would not be so unless it were in potency with respect to some nobler form and in some way perfectible by it. And this can also be established by the fact that the intellective is not unible to the sensitive power in animals other than man, because the sensitive power in them has no potency for the reception or possession of the intellective.[11]

Adam thus gives us a definite, if hesitant, answer to the question of whether the vegetative and sensitive souls are of the same substance as the intellective: his answer is no, at least according to Aristotelian principles. His attempt to show how this could be, even though it seems to be inspired by Philip the Chancellor, is more cogently stated than similar material in Philip's *Summa*, but Adam gives proportionately less attention to it. Nor does he mention Philip's insistence that body and soul must be united by some medium.

RICHARD FISHACRE. Richard Fishacre was the first to hold the second Dominican chair at Oxford and the author of the first commentary on the *Sentences* to come out of the Oxford schools.[12] He composed this commentary between 1241 and 1245 and so was contemporary with John of La Rochelle, Albert the Great, Peter of Spain, and Roger Bacon at Paris. Although he was strongly influenced by both the works and the personality of

---

[11] "Ad quartum potest dici quod seipsis uniuntur et non indigent alio uniente, sicut forma et materia: est enim anima intellectiva actus et perfeccio aliquo modo sensitive, et sensitiva vegetative. Illud enim quod prius est inter animas est in potencia respectu persterioris et aliquo modo perfectibile ab ipso. Et quod virtus sensitiva sit in potencia respectu intellective patet per hoc quod ymaginacio, que est ultima virtus sensitiva, perfeccior est et magis determinata in homine quam in aliis animalibus: quod non esset nisi esset in potencia respectu forme nobilioris et aliquo modo perfectibilis ab ipsa. Ex hoc etiam potest haberi propter quid non est intellectiva unibilis virtuti sensitive in aliis animalibus ab homine, quia virtus sensitiva in ipsis nullam habet potenciam ad recepcionem intellective nec habilitatem." Callus, *art. cit.*, p. 438

[12] This commentary is currently being edited by an international group of scholars under the general editorship of R. James Long. Until this edition appears, the most important work on Fishacre's doctrine of the soul is R. James Long, *The Problem of the Soul in Richard Fishacre's Commentary on the Sentences* (Unpublished doctoral dissertation, University of Toronto, 1968).

Grosseteste,[13] he was often at odds with him and did his best to keep his theological teaching abreast of the times. He was not a brilliant or innovative thinker, but rather a thoroughly competent and conscientious provincial theologian.

The influence of Grosseteste is evident in several matters. Although, like the vast majority of thirteenth-century theologians, Fishacre considered the body and soul to be distinct substances, he followed Grosseteste in denying that man is essentially a soul.[14] His emphasis on the soul's strong desire to be united to the body seems also to be indebted to Grosseteste. In fact, says Richard, so strong is this desire that if the soul were eternally deprived of its body, it would be eternally miserable, and even if Adam's soul had had foreknowledge of its fall and reparation, it would still have chosen to enter the body.[15] He also followed Grosseteste in positing light as the medium between body and soul, but whereas Grosseteste had emphasized light as the means by which the soul moved the body, Fishacre describes it as the chain (*vinculum*) uniting the two. From Grosseteste too he took the notion of 'incorporation.' Grosseteste had used this notion in his scientific works, particularly *De calore solis*,[16] but not in his discussions of the relationship of body and soul. But Fishacre seems to have envisioned the soul as something like the popular notion of a ghost. Although as a substance it occupied only a point, nevertheless when it was joined to a body, its spatial extension and shape exactly matched that of the body. It was incorporated in the body in the same way as Grosseteste had the sun's rays incorporated in the air—spatially congruent but substantially distinct. As a result of its incorporation, the soul was also subject to mutability, the extent of which depended on the mutability of the body in which it was incorporated, although considered in itself it was immutable.

---

[13] See Richard C. Dales, "The Influence of Grosseteste's *Hexaemeron* on the *Sentences* Commentaries of Richard Rufus, O.F.M. and Richard Fishacre, O.P.," *Viator* 2 (1971), 271-300.

[14] R. James Long, "Richard Fishacre and the Problem of the Soul," *The Modern Schoolman* 52 (1975), 263-70, on p. 70.

[15] "Sed cum naturaliter habeat ad corpus inclinationem: si aeternaliter careret corpore, aeternaliter careret suo appetibili. Igitur aeternaliter esset misera, si non intraret corpus. ... Quid etiam impediat animam Adae praescivisse, antequam intrasset corpus, et lapsum et reparationem, et tamen affectasse intrare corpus? Videtur enim quod, etiam si scivisset lapsum et reparationem, libentius eligeret intrare corpus quam aeternaliter carere corpore, ex cuius carentia esset aeternaliter misera." Printed in Long, "Richard Fishacre," p. 268, n. 24.

[16] See Richard C. Dales, "Robert Grosseteste's Scientific Works," *Isis* 52 (1961), 381-402, on pp. 398-99.

But Fishacre was heir to many other influences. He followed Avicenna in holding that soul and body are distinct substances, and that the soul must be distinguished as to its operations in the body, when it functions as the form or perfection of the body, and as it is in itself, a complete substance. He also accepted universal hylomorphism without question. His view of matter is quite un-Aristotelian; he seems to consider it primarily as dross. The more matter a thing has, the less noble it is, and the less matter the more noble. Since the human soul is the noblest of forms, it has the least matter. He also explicitly admits a plurality of forms in the soul, not in the sense of the vegetative, sensitive, and intellective souls being distinct forms, but as an explanation of the two faces of the soul. It has one form in common with the angels, which constitutes it as a rational spiritual being, and another form, subordinated to this, by which it is properly a soul,[17] which he later identifies as the sensitive soul.[18]

On the question of the unity or plurality of forms in the soul with respect to the vegetative, sensitive, and intellective functions, Fishacre, like his Oxford colleagues, is indecisive. For the purposes of discussion, he distinguishes three proposed solutions to the problem, but he rebuts each one with what he considers to be an insuperable objection."Some people," he says,"think that the vegetable, sensible, and rational are one and the same form and vary only according to operation, just as the sensible soul is one form having many operations, such as sight, hearing, and things of this sort."

But there is a serious objection to this, because

> every action has one formal and substantial cause, and
> every substantial form has a single action. Therefore, if
> these actions differ essentially, they will either be actions of
> forms differing essentially, or at least of organs differing
> essentially. Therefore, if sense, nutrition, and under-
> standing are essentially diverse operations, it is clear that,
> since the organs do not differ, the forms will differ
> essentially from those things by which they are caused. But
> it is not the same with sight and hearing. Indeed, these
> operations do not differ essentially, but only with respect to
> different instruments. For the soul acts the same way on
> the eye when the eye sees as it does on the ear when the ear
> hears. The one action does not differ from the other except

---

[17] Anima enim aliquo sui convenit cum angelo et aliquo sui differt. Habet enim anima formam unam partem sui quae est communis ipsi et angelo qua est spiritus rationalis; et etiam formam habet sibi propriam, qua scilicet habet affectionem ad corpus, et qua proprie est anima." Long, "Richard Fishacre," p. 264, n. 8.

[18] Long, "Richard Fishacre," p. 265.

as liquefaction and congealing, which are one action of the sun.[19]

Then he presents a somewhat inexact account of Philip the Chancellor's view:

> Others posit that in man the soul is numerically one form, but it has other forms reciprocally ordered; and the act of vegetation comes forth from one form, the act of sensing from another, and from the third the act that is understanding, so that the form closest to matter is that from which the vegetative comes about, and the form superadded to it is a nobler form, from which sensing arises, and the noblest, from which is understanding, so that with respect to the noblest form, the preceding ones are like material dispositions, and they seem to be such because that which is a genus predicates only the form from which sense comes forth, but the rational, like a differentia of the same genus, [predicates] a nobler form superadded to it.[20]

But there is also a serious objection to this view. To appreciate this objection, we must remember that Richard considers the soul to share the form of

---

[19] "Estimant enim aliqui, quod vegetabilis et sensibilis et rationalis sunt una et eadem forma, et variantur tantum secundum operationem: sicut anima sensibilis est unica forma habens multas operationes, scilicet videre, audire, et huiusmodi.—Contra quod merito sic opponitur. Omnis actio una causam habet unam formalem et substantialem, et omnis forma substantialis unicam habet actionem. Igitur, si sunt actiones essentialiter differentes, aut erunt actiones formarum essentialiter differentium, aut ad minus organorum essentialiter differentium. Igitur, si sentire, vegetari, intelligere sunt essentialiter diversae operationes, patet quod, cum organa non differant, erunt formae essentialiter differentes a quibus causantur. Non est autem simile de eo quod est videre et audire: quippe istae operationes non differeunt essentialiter, sed tantum ratione instrumentorum diversorum. Idem enim agit anima in oculum, cum videt oculus, quod agit in aurem cum audit auris; nec differt haec actio ab illa nisi sicut liquefacere et constringere quae sunt una actio solis." Printed in Raymond-M. Martin, O.P., "La Question de l'Unité de la Forme substantielle dans le premier Collège dominicain à Oxford (1221-1248)," *Revue néoscholastique de philosophie* 22 (1920), 107-12, on p. 110.

[20] "Propterea, alii posuerunt quod in homine est anima unica forma numero, habens tamen formas invicem ordinatas diversas; et ab una forma egreditur actus vegetationis, ab alia actus quod est sentire, a tertia actus quod est intelligere; ut forma proximior materiae sit illa a qua est vegetatio, et forma ei superaddita sit forma nobilior, scilicet a qua est sentire, et nobilissima a qua est intelligere; ut respectu hujus nobilissimae sint duae formae praecedentes, quasi dispositiones materiales, et talem quidem habere videntur habitudinem quia illud quod est genus non praedicat nisi formam a qua est sensus, rationale vero tanquam eiusdem generis differentia, formam nobiliorem ei superadditam." Martin, *art. cit.*, pp. 110-11.

rationality with angels. It would seem, he says, according to the view just presented, that the vegetative and sensitive forms must precede in order for the rational to come about. "But in angels the vegetable and sensible do not precede the intellectual. Therefore these forms do not behave as material dispositions toward the form by which understanding arises."[21]

The third solution he presents is that "there are three forms and three *hoc aliquid* in man, by which these three operations exist. But there are not thereby three souls of man, but one soul, consisting of three substances that differ substantially, just as one hand consists of nerves, bones, and flesh, which are essentially different."[22] This view, he says, is plainly contradicted by Augustine in the book *De definitionibus recte fidei*, chapter 15, which he then quotes.

And he concludes: "But which of these three opinions is truer, I do not dare to define. The word soul therefore can here bespeak something common to the vegetable, sensible, and rational according to the third opinion; or one substance of the soul according to the first and second opinions."[23]

RICHARD RUFUS OF CORNWALL. Fishacre died in 1248. By this time Richard Rufus of Cornwall, who had just completed his degree in arts at Paris, was pursuing a degree in theology at Oxford. He had joined the Franciscan order shortly before returning to England in 1238, and his *Sentences* commentary, composed between 1246 and 1250, was the first to come out of the Oxford Franciscan school.[24] Apparently emotionally distraught when the Oxford Franciscans promoted Thomas of York over him to be lecturer to the convent in 1253, he took advantage of an earlier permission to resume his studies at

---

[21] "Sed in angelis non est vegetabilis vel sensibilis praecedens in eis formam a qua est intelligere. Igitur non sunt haec formae se habentes ad formam a qua est intelligere ut dispositiones materiales." *Ibid.*, p. 111.

[22] "... sunt tres formae et tria haec aliquid in hominibus a quibus sunt istae tres operationes; nec propter hoc sunt tres animae hominis, sed una anima constans ex tribus substantiis essentialiter differentibus: sicut una est manus constans ex nervis, ossibus et carne quae essentialiter differunt." *Loc. cit.*

[23] "Quae autem harum trium opinionum verior sit, diffinire non audeo. Li *animae* igitur hic dicere potest quiddam commune dictis tribus, scilicet vegetabili, sensibili, rationabili, secundum tertiam opinionem; vel ipsam substantiam unam animae, secundum primam et secundam opinionem." *Loc. cit.*

[24] See F. Pelster, "Der älteste Sentenzenkommentar aus der Oxforder Franziskanerschule," *Scholastik* 1 (1926), 50-80. On the problem of the date, provenance, and authenticity of the commentary, see Peter Raedts, *Richard Rufus of Cornwall and the Tradition of Oxford Theology* (Oxford, 1987), pp. 20-39.

Paris, where he made his abbreviation of the lectures of his predecessor, Bonaventure. In 1256 he returned to Oxford to be the fifth lecturer to the Franciscan convent, his *bête noir* Thomas of York having been transferred to Cambridge.

Although Rufus made extensive use of the writings of his two most important recent predecessors at Oxford, Robert Grosseteste and Richard Fishacre,[25] he disagreed on important points with both men. One of Grosseteste's most unusual positions was that the souls of both men and beasts were created directly by God, not generated by natural processes, and were infused at the moment of conception. Fishacre had copied these remarks from Grosseteste's *Hexaemeron* without questioning them. But Rufus considers the position theologically untenable.[26] He has little interest in Grosseteste's lengthy discussion of man as the image of God, but he does agree with him that it is the whole man, and not just the soul, that constitutes this divine image.[27] He also shares Grosseteste's emphasis (as had Fishacre) on the powerful natural desire of the soul to be united to a body and to serve it.[28]

He has a fundamental disagreement with Fishacre over whether the soul shares any changeable principle with the body; Rufus says it does not. And as to whether the human soul is composed of matter and form, Rufus says that it is, and this is the reason it can subsist apart from the body. But animal souls are purely material forms; their matter is the body of which they are forms.[29] So while he accepts the matter-form composition of the human soul, he does not explicitly accept either universal hylomorphism or the plurality of forms. He also argues that since what is corrupted must have been generated, and since animal souls are corrupted with the body, they must therefore have been generated. But human souls, since they can subsist apart from the body, were created rather than generated.[30]

---

[25] See Richard C. Dales, "The Influence of Grosseteste's *Hexaemeron*," and Raedts, *Richard Rufus*, passim.

[26] For a full discussion of this point, see Raedts, *Richard Rufus*, p. 163.

[27] Balliol College MS 54, fol. 142ra, printed in Raedts, *Richard Rufus*, p. 170. Cf. Grosseteste, *Hexaemeron* VIII, i, 3 and VIII, ii, 1, ed. Dales and Gieben, pp. 227, 230-31. See also Dales, "The Influence," pp. 287-88.

[28] See Raedts, *Richard Rufus*, p. 172.

[29] See the discussion in Raedts, *Richard Rufus*, p. 164.

[30] *Loc. cit.*

Rufus then was willing to take some definite positions concerning the soul, and in doing so he had had the courage to disagree with the opinionated bishop of Lincoln and patron of the mendicant orders, Robert Grosseteste. But on the question of how the vegetative, sensitive, and rational functions are related in the human soul, he was unable to come to a conclusion. But it was still a question that had to be raised, even if not answered, and in book 2, distinction 7, chapters 167-70 of his commentary on the *Sentences*, he conducts a lengthy, tortuous, and inconclusive examination of the problem.[31]

Rufus begins his discussion by setting the view of Augustine (actually Alcher of Clairvaux) *De spiritu et anima* and that of the philosophers in sharp contrast to each other, quoting sections from *De spiritu et anima* that unequivocally state that the simple human soul performs all the life functions, and then presenting five arguments for the view that the rational soul is of a different substance from the vegetative and sensitive. But at this point, instead of proceeding to take one side of the other and determining the question, he writes: "But what can I say?" There seems to be an irreducible difference of opinion between the theologians and the philosophers, and Rufus figuratively ties himself in knots in presenting arguments and counterarguments on each side, and then again seems to throw up his hands:

> What can I say about these things? The philosophers say
> that there are three substances but one soul in man, because
> the vegetative is in potency to the sensitive, and it to the
> intellective; and the intellective is the perfection, and 'soul'
> is the name of the perfection. ... And according to this, the
> rational alone is called the soul in man, but it is made up of
> three substances. And to assist in understanding how one is
> in the other as if it were matter and nevertheless one soul is
> made from them, they give the example of the Philosopher
> about figures and numbers, and the example of the ray of
> fire and ray of the sun, which unite to make one ray; and
> that of fire is corrupted while that of the sun remains. Thus
> two things are corrupted with the body and the third

---

[31] This section of the commentary has been edited by Callus, "Two Early Oxford Masters," pp. 439-45. Fr. Callus has identified Richard's sources for this section as : Philip the Chancellor, *Summa de bono*, John of La Rochelle, *Summa de anima*, Roland of Cremona, *Comm. in Sent.*, Augustine, *De Genesi ad litteram*, Alcher of Clairvaux, *De spiritu et anima* (ascribed to Augustine by Richard), Grosseteste, *De statu causarum*, Nicolaus of Damascus, *De plantis* (ascribed to Aristotle), Hugh of St. Victor, *Explanatio in Cant. B.M.V.*, Fishacre, *Comm. in Sent.*, as well as Aristotle, *De anima* and *De animalibus*.

remains, and it is separated from the body like the perpetual
from the corruptible.[32]

At this point, like Adam of Buckfield and Richard Fishacre, Rufus arranges
the views into three classes. The first class accepts the position he has just
been discussing (i.e., of Philip the Chancellor), and those who hold it do not
care much about the authority of the saints, who are only inspired by the Holy
Spirit when they are speaking about matters pertaining to faith, but in natural
philosophy they speak only as men.

The second group accepts the view of the theologians completely, that is,
that the soul is one substance with three powers. The aspect of this position
that bothers Richard most is that its proponents concede that the vegetative
and sensitive powers in man and beast should be understood univocally; but
in beasts they perish with the body, while in man they remain, for they will be
needed in some fashion at the resurrection of the body. After noting some
arguments to account for this, including Roland of Cremona's doctrine that
the embryo grows by virtue of the mother's vital powers, not its own, Richard
mentions a position that attempts to resolve the conflict between philosophers
and theologians. With the theologians, its proponents admit a single soul,
coming from without, with all three powers. But in the embryo there are also
vegetative and sensitive powers that come forth from matter. By these latter,
man is made an incomplete animal capable of being perfected by the rational
soul with its potencies, so that it might be a completed animal in the species
'man.' And, he notes, these are not substances, but only powers.

He then suggests that this compromise position might be strengthened by
something he had read at Paris in a book of Hugh of St. Victor (i.e.,
Explanatio in canticis Beatae Mariae Virginis). Hugh, he says, seems to
agree with the third opinion, that before the infusion of the rational soul there
is some power that is a principle of growth and motion, and so there is some
sort of life in the embryo, but it is not complete in species (that is, it is not a
human life). Hugh, he goes on, does not seem to contradict those who hold
this, except insofar as they posit in the embryo an irrational soul through

---

[32] "Quid in hiis dicam? Philosophi dicunt quod sunt tres substacie, set una anima in
homine, scilicet quia vegetativa est possibilis respectu sensitive, et hec respectu intellective,
illa autem est perfeccio, et anima nomen est perfeccionis. ... Et secundum hoc in homine sola
racionalis dicitur anima, tamen tres substancie. Et ad intelligendum quomodo una est in altera
quasi materialis, et tamen sit una anima in illis, est exemplum Philosophi in figuris et numeris.
Et ad exemplum datur quod radius solis et ignis simul uniuntur, et contingit quod ignis
corrumpitur et remanet solaris. Sic due corrumpuntur cum corpore et tercia remanet, et
separatur a corpore sicut perpetuum a corruptibili." Callus, "Two Early Oxford Masters,: pp.
441-42.

which a completed animal lives in the species 'beast,' although the rational soul has not yet been infused.

Richard seems to favor this explanation, with the correction of Hugh of St. Victor, but he still refuses to take a stand, and he concludes this long section with: "Ecce in hiis omnibus non diffinio." He had a genius for detecting flaws in the arguments of others, but he was not nearly so good at constructing his own. And most characteristic of him perhaps was that he was quite willing to stress difficulties and leave many questions unanswered.

R. DE STANINGTONA. Probably during the mid-1250s, an Oxford friar (in all likelihood a Dominican), known to us only as R. de Staningtona, composed an admirable summary of Aristotle's *libri naturales* entitled *Compilacio quedam librorum naturalium*, contained on folios 100ra through 125va of Bodleian manuscript Digby 204. It has never been edited, although I have described it and printed selections from it in the *Journal of the History of Philosophy*.[33]

The last work to be summarized is *De anima*, occupying folios 121ra-122ra.

This is for the most part a clear summary of the contents of *De anima*, assisted by the explanations of Averroes, but at the top of folio 121rb the summary is interrupted by a *quaestio*: "Utrum in anima sit eadem substancia vegetative, sensitive, et intellective, an sint diverse substancie." This *quaestio* is based upon Fishacre's commentary on the *Sentences*, and from the phrase: "Ad hoc dicendum quod triplex est opinio," on folio 121rb, the correspondence is almost verbatim. The question ends with the same quotation from Gennadius's *De ecclesiasticis dogmatibus*, cited as "Augustinus in libro de diffinicionibus recte fidei;" and Stanington, like Fishacre, and in almost the same words, declines to take a position: "Que autem horum opinionum verior sit diffinire non sufficio." One would like to know how Stanington would have treated book 3, chapter 5, but his summary breaks off in mid-sentence before it reaches this point.

Although this work possesses no independent philosophical or theological significance, it was most likely used in the Oxford Dominican convent for the preliminary education of the brothers, many of whom were not degree candidates, and so it provided a medium for Fishacre's influence among those who were not likely to read his *Sentences* commentary.

ROBERT KILWARDBY. Shortly after Stanington composed this work, the Oxford Dominican convent received a young master who for the preceding fourteen years had been studying and teaching arts at Paris. This was master

---

[33] Richard C. Dales, "R. de Staningtona: An Unknown Writer of the Thirteenth Century," *Journal of the History of Philosophy* 4 (1966), 199-208.

Robert Kilwardby, who had just recently assumed the Dominican habit. Although very few of Kilwardby's philosophical works have been edited, he is credited by Nicolas Trivet and the Stams catalogue with commentaries on nearly the entire corpus of Aristotle's natural philosophy. He was regent master in arts at Paris from about 1237 to 1245, after which he returned to England and entered the Dominican order in or shortly before 1250. He took his degree in theology at Oxford and succeeded (probably) Peter of Manners as Dominican regent in theology. He was elected prior provincial for the English province in 1261, a position he held until 1272, when he was elected archbishop of Canterbury. In this latter capacity, on March 18, 1277, he issued a list of propositions in grammar, logic, and natural philosophy, which he forbade being taught at Oxford. It has usually been assumed that this was directed against Thomas Aquinas, although Leland Wilshire has ably questioned this interpretation.[34] In 1278 he was named cardinal-bishop of Porto and repaired to Rome. He died at Viterbo in the following year.

Although Kilwardby's questions on book 2 of the *Sentences*, in which his views of the soul would be most fully elaborated, is not yet available in print,[35] the main features of his position can be gleaned from an examination of his *De ortu scientiarum*,[36] written shortly after his entry into the Dominican order at an elementary level for beginning students, and from his letter to Peter of Conflans, defending himself for having forbidden the teaching of views that had been taught by his fellow-Dominican Thomas Aquinas and were currently being maintained at Paris. According to Trivet, after Kilwardby became a Dominican he ceased his scientific and philosophical studies to devote himself to Scripture and the Fathers.[37] It is not unreasonable to suppose that he was largely unaware of philosophical developments during the decades when he had been absent from the schools.

---

[34] Leland E. Wilshire, "Was the Oxford Condemnation Directed Against Aquinas?" *The New Scholasticism* 48 (1974), 25-32 and "The Oxford Condemnation of 1277 and the Intellectual Life of the Thirteenth-Century Universities," in *Aspectus et Affectus. Essays and Editions in Grosseteste and Medieval Intellectual Life in Honor of Richard C. Dales*, ed. Gunar Freibergs (New York, 1993), 112-24.

[35] Those portions that have appeared are: *Quaestiones in librum primum Sententiarum*, ed. Johannes Schneider (Munich, 1986); *Quaestiones in librum tertium Sententiarum I. Christologie*, ed. Elizabeth Gössman (Munich, 1982); and *Quaestiones in librum tertium Sententiarum II. Tugendlehre*, ed. Gerhard Liebold (Munich, 1985).

[36] Robert Kilwardby, O.P., *De ortu scientiarum*, ed. Albert G. Judy, O.P. Auctores Brittanici Medii Aevi 4 (London, 1976). I have summarized the biographical information in the preceding paragraph from pp. xi-xvii of this work.

[37] *Annales Sex Regum Angliae*, ed. A. Hall (Oxford, 1719) I, p. 235.

His explanation of his own views on the soul sounds very much like those
that were current at Paris during the 1240s. It is clear throughout his
philosophical writings that he was much influenced by Robert Grosseteste,
and in his *De ortu scientiarum* he paraphrases Grosseteste's view on the
soul's strong desire to be united to the body: "... the human soul naturally
desires to be united to the body and naturally hates to be sundered from it."[38]
But he was writing and teaching in a much more Aristotelian atmosphere. He
did not, like Grosseteste, ignore Aristotle's works on the soul but used them
extensively and without trepidation, and indeed he praised them highly.[39] He
makes a clear distinction between the way a physicist must treat the soul (that
is, as it is joined to the body) and the way a metaphysician treats it (i.e., as a
separate substance in its own right).[40] He opts for the matter-form
composition of the soul, distinguishing three kinds of matter: the first, when it
is united to form, causes number; the second confers magnitude; and the third
supports all the kinds of accidents that fall under the consideration of the
natural philosopher. The first kind is superior in status according to nature,
but the last is prior in human knowledge and is the meaning it almost always
has in the philosophy of Aristotle and the commentaries of Averroes.[41]

The difficulty of speaking of 'unity of form' without further qualification is
made apparent in Kilwardby's views on the unity of the soul. In *De ortu
scientiarum* he had included a discussion of how parts may be predicated of
wholes,[42] and in his letter to Peter of Conflans, denying that he had forbidden
"the position of the unity of forms," he gives an account of his own view,
which he considers to be the universally held position in philosophy:

> Under these words [i.e., positio de unitate formarum], such
> an article has not been prohibited at Oxford, nor do I recall
> having heard of it. Why they call it the position of the unity
> of forms, I do not sufficiently understand. I know many
> corporeal and spiritual forms that have no unity among

---

[38] "... humana anima naturaliter appetit corpori uniri et naturaliter odit dissolvi ab
eo." *De ortu scientiarum* 37, 358, *ed. cit.*, p. 127.

[39] "... tractatus *De anima* et libri ad eius evidentiam sequentes, scilicet *De sensu et
sensato, De somno et vigilia, De morte et vita*, in quorum primo, scilicet *De anima*, destruit
opiniones falsas antiquas de anima et suam veritatem de ipsa tradit." *De ortu scientiarum* 10,
48, *ed. cit.*, p. 25.

[40] *De ortu scientiarum* 28, 233, *ed. cit.*, p. 88.

[41] *De ortu scientiarum* 31, 320-21, *ed. cit.*, p. 114.

[42] *De ortu scientiarum* 31, 316, *ed. cit.*, pp. 112-13.

themselves; I also know some that have some degree of
unity, but not in every respect; and I know that a diversity
of action proceeds from a diversity of forms; ... but what
complete unity there may be of forms that exercize diverse
actions with respect to species and require diverse objects
with respect to species, I neither know nor understand, but I
judge it to be false and impossible. But when one speaks
about the unity of forms as meaning that when the last form
to arrive performs the actions of all the preceding ones by
itself, such a position is intolerable and false. For we see
sensibly in a man flesh, bone, nerves, blood, eye, foot, and
such like, of which none is without its true and proper
form.[43]

Then, after a string of arguments showing why such a position is contrary to
sound philosophy and subversive of the Christian faith, Kilwardby
investigates its application to the human soul:

Unless more be said, I do not sufficiently understand what
"positio de unitate formarum" means. But I know that one
man has one form, which is not simple but composed of
many [forms] having a natural order to each other and
without any one of which a man would not be able to be
complete. The last of these, which completes and perfects
the whole aggregate, is the intellect. For just as, from many
differentiae naturally ordered to each other, there is one
form of the composite, thus it is also in things composed by
nature from the forms constituting them. And just as, on
the part of the body, there are many members, each of
which has its own proper form and proper matter, of which
none is the other, and they nevertheless constitute one body

---

[43] "Sub his quidem verbis articulus iste non erat Oxonie prohibitus, nec illum
memini me audivisse. Quare autem dicitur "positio de unitate formarum" non satis intelligo.
Novi enim multas formas corporales et spirituales, que nullam habent ad invicem unitatem;
novi etiam aliquas quandam habere unitatem, sed non omnimodam; et novi a diversitate
formarum diversitatem procedere actionum ...; sed quid unitas sit formarum omnimoda,
diveras specie actiones exercentium et diversa specie obiecta requirentium, nec novi nec
intelligo, sed falsum iudico et impossibile. Exponitur autem quod predictum est de unitate
formarum in hunc modum, quod ultima forma adveniente, que est perfectiva compositi,
omnes alie que precesserunt citra materiam currumpuntur et ultima adveniens per ipsam agit
omnium actiones. Istud intolerabile est et impossibile et primo quod ad sensum videmus.
Videmus enim sensibiliter in homine carnem, os, nervum, sanguinem, oculum, pedem et talia,
quorum nullum est sine vera et propria sua forma." A. Birkenmajer, "Der Brief Robert
Kilwardbys an Peter von Conflans und die Streitschrift des Ägidius von Lessines," BGPM
20.5, pp. 46-69, on p. 60.

> through a natural order and connection that they have
> toward each other, but they do not constitute one *simple*
> body—thus, on the part of the soul, there are many parts
> that differ essentially but that nevertheless make one soul,
> not in such a way that the soul is simple through its essence
> but is one form of a living thing. Any other unity of forms
> in composite beings neither does philosophy teach nor do I
> grasp.[44]

These words were written nearly thirty years after the *De ortu scientiarum*, and much had happened in the meantime, of which Kilwardby seemed unaware. His horrified reaction to Aquinas's position, while he still claimed not to oppose the 'positio de unitate formarum,' is that of an Aristotelian (certainly not an Augustinian) of the 1240s suddenly and without preparation being confronted by a radically different notion of form, matter, privation, and unity, than those current when he was a regent master in arts. It is to the Parisian milieu of the 1240s, in which Kilwardby's view of the soul was formed, that we now turn.

---

[44] "...'Positio de unitate formarum,' nisi plus dicatur, non satis est mihi. Scio tamen, quod unus homo unam formam habet, que non est una simplex, sed ex multis composita, ordinem ad invicem habentibus naturalem et sine quarum nulla perfectus homo esse potest, quarum ultima, completiva et perfectiva totius aggregati est intellectus. Sicut enim ex multis differentiis ad invicem naturaliter ordinatis una diffiniti sit forma, sic est in rebus compositis per naturam de formis constituentibus eas; et sicut ex parte corporis multa sunt membra proprias formas et propriam materiam habentia, quorum nullum est alterum, tamen constituunt unum corpus per ordinem et colligationem naturalem, quam habent ad invicem, sed non constituunt unum corpus simplex—sic ex parte anime sunt multe partes essentialiter differentes, que tamen per ordinem et colligationem naturalem unam animam efficiunt, non tamen ita, quod anima sit simplex per essentiam, sed una forma viventis, et ex his formis corporalibus iam memoratis et hac spirituali, que constat ex multis, humanitas una resultat. Aliam unitatem formarum in compositis nec vult philosophia nec ego concipio." *Ibid.*, pp. 63-64.

CHAPTER FOUR

TWENTY YEARS OF CONFUSION: PARIS

It was during the 1240s that Aristotle's *libri naturales* came to be regularly lectured on at Paris (and, as we surmise from the commentaries of Adam of Buckfield, also at Oxford), although they were not yet *ad formam* books. Their use was facilitated by the recently acquired commentaries of Averroes, which greatly aided the scholastics' comprehension of the Stagyrite's thought. However, it took over two decades for the full import of Averroes's interpretation of Aristotle's text to be fully appreciated. That Averroes claimed that Aristotle had taught the world to be beginningless was realized immediately, although most scholastics considered the Commentator to be wrong about this, and so denied that Aristotle had taught the doctrine in a way incompatible with Christianity. It was otherwise however with Averroes's interpretation of Aristotle's teaching on the agent and possible intellects, and, as Dominique Salman has convincingly shown, until the decade of the 1250s, Averroes was considered to have held that the intellect was a part of the individual human soul.

In spite of the help offered by Averroes, however, other factors prevented an accurate comprehension of Aristotle's meaning, especially: the number of works falsely ascribed to him (particularly the *Liber de causis*); and, more importantly, the overweaning influence of Avicenna. Among the writers we shall discuss in this chapter, all, whether artists or theologians, were dominated by the thought of Avicenna and particularly, for the topic we are investigating, his *Liber sextus naturalium*. This influence operated differently with different writers. None followed him in all things, but none escaped his influence.

A. ARTISTS

PETER OF SPAIN. Peter of Spain was known in his own day as a great physician. Today, aside from the fact that he was pope as John XXI in 1276 and 1277, he is known primarily as the author of the widely used textbook *Summulae logicales*. But it is often overlooked that, along with Albert the Great, he was one of the leaders in introducing Aristotle's natural philosophy to Latin Europe. He composed commentaries on a considerable portion of

the Aristotelian corpus and was especially interested in the *parva naturalia*.[1]
Although Peter began his scholastic career at Paris, he spent much of it
elsewhere.   It is therefore somewhat arbitrary, though organizationally
expedient, to include him among the Parisian artists. The principal facts of
his biography have been established with a high degree of probability by L.
M. De Rijk.[2] He was born at Lisbon around 1205 and studied and taught arts
(and possibly began the study of theology) at Paris from 1220 until the
suspension of lectures in 1229. In that year he returned to Spain (probably
Léon), and in 1235 came north of the Pyrennes, first to Toulouse and then to
Montpellier, where he studied and perhaps taught medicine. From 1245 to
1250 he taught medicine at Sienna, and in the latter year returned to Portugal.
In 1263 he was named *magister scholarum* at Lisbon. In 1272 he became
personal physician to Gregory X, and the next year became successively
archbishop of Braga and cardinal-bishop of Frascati. He was elected pope on
September 15, 1276 and died on May 20, 1277.

Since Peter was pope in 1276, when complaints of dangerous novelties being
taught at Paris arrived in Rome, and since, as John XXI, he ordered the bishop
of Paris to look into the matter, it is of more than ordinary interest to
determine his own doctrine of the soul. This is especially true because it has
been suggested that Peter instigated the condemnation in order to quash views
contrary to his own.[3] This task is more difficult than it would seem at first
because there are three separate works on the soul attributed to Peter, and they

---

[1]   A complete bibliography of Peter's works is printed in P. Manuel Alonso, ed.,
Pedro Hispano Obras Filosóficas, II, Comentario al "De anima" de Aristóteles (Madrid,
1944), pp. 14-27.

[2]   L. M. De Rijk, "On the Life of Peter of Spain, the Author of the Tractatus
Called Afterwards *Summulae logicales*," *Vivarium* 8 (1970), 123-54.

[3]   This was first argued by André Callebaut, O.F.M., "Jean Pecham O.F.M. et
l'augustinisme. Aperçus historiques (1263-1265)," *Archivum Franciscanum Historicum* 18
(1925), 441-72.   It was repeated by Martin Grabmann, "Mittelalterliche lateinische
Aristotelesübersetzungen   und   Aristoteleskommentare   in   Handschriften   spanischer
Bibliotheken," *Sitzungsberichte der Bayerischen Akademie der Wissenschaften*. Philologisch-
philosophische Klasse (Munich, 1928), 98-113 and "Die Lehre vom Intellectus Possibilis und
Intellectus Agens im Liber De Anima des Petrus Hispanus des späteren Papstes Johannes
XXI," *AHDLMA* 12-13 (1937-38), 167-208, who remarked (pp. 180-81) that the greater part
of Aquinas's argument against Avicenna in *Summa contra Gentiles* II, 74-78, could just as
well be applied against Peter. This judgment was also accepted by Gilson, "Les sources
gréco-arabes de l'augustinisme avicennisant," *AHDLMA* 4 (1929-30),5-149, on pp.106-07.

do not agree on all crucial points. One of these, the *Scientia libri de anima*,[4] first noticed by Martin Grabmann,[5] is contained in Madrid, Bibl. nac. MS 3314, folios 3r-67v and is almost certainly authentic, since it concludes with the less than modest signature:

> Ego igitur Petrus hispanus Portugalensis liberalium artium doctor, phylosophice sublimitatis gubernator, medicinalis facultatis decor ac proficue rector, in scientia de anima decrevi hoc opus precipuum componendum, pro cuius complemento divine bonitatis largitas gratiarum actionibus exaltetur.[6]

This work is not a commentary, nor does it consistently use the *disputatio* format to present its teaching. It is an independent treatise, quite free in form, and it cites no authorities at all. However, it treats only those matters covered in the latter part of book 2 and book 3 of Aristotle's *De anima*.

The second, entitled *Expositio libri De anima*[7] is essentially concerned with dividing the text, elucidating the structure of the argumentation, and making clear Aristotle's doctrine, but it also contains occasional *dubia*. Although it examines *De anima* 3, 3-5 with great care, there is no hint that the author is aware of the doctrine of monopsychism, although Averroes is cited from time to time.

The third work on the soul attributed to him is a commentary on *De anima* contained in Cracow, Univ. MS 726, folios 41r-134r,[8] and it is also incomplete. This work is also explicitly ascribed to Peter of Spain, and after a lengthy discussion of the question, its editor, Manuel Alonso, judged it to be authentic.[9] However, Étienne Gilson has expressed serious doubts concerning

---

[4] P. Manuel Alonso, S.I., ed., *Scientia Libri De Anima por Pedro Hispano* (Madrid, 1941).

[5] Martin Grabmann, "Mittelalterliche lateinische Aristotelesübersetzungen."

[6] *Ed. cit.*, p. 564.

[7] *Pedro Hispano Obras Filosóficas, III, Expositio libri De anima. -de morte et vita et De causis longitudinis et brevitatis vitae. -Liber naturalis de rebus principalibus*, ed. Manuel Alonso, S.I, (Madrid, 1952).

[8] See above, n. 1.

[9] *Ed. cit.*, pp. 33-50.

its authenticity.[10]   Although there are differences of style between it and the
two other works and one important difference in doctrine, I shall nevertheless
accept the judgment of the editor in considering it to be an authentic work of
Peter of Spain.

It is difficult to determine the composition dates of Peter's works on the soul.
It seems clear however that the commentary on *De anima* came first. Its form
ties it clearly to the schools—in fact, it probably represents a lecture course,
although whether at Paris or elsewhere is not certain. The *Expositio*, also
apparently a lecture course, seems to have come next, since in it the author
alludes to his having treated the earlier part of *De anima* previously. The
Scientia libri de anima, on the other hand, is not 'scholastic' in its format. It
seems to have been written by a confident, mature scholar, secure in his
position as an authority. We know that he was rector of the medical faculty at
the time he composed it, but we do not know at which university. We might
date the first two works as around 1240 and the *Scientia* about a decade later.

Peter's thought is an interesting congeries of ideas from many sources,
expertly but not perfectly fitted together and modified by his own mind. In
both works, Peter, like Albert the Great, follows Avicenna in teaching that the
soul must be considered from two points of view: in itself, and in its relation
to the body. In itself it is a substance participating in reason; with respect to
the body (and it was only thus that Aristotle's definition applied to it), it is the
formal principle, or act, that perfects the body.[11]   And although Peter
discussed it from both perspectives, he was much more interested in the way it
operates in conjunction with the body.

It is in the discussion of the soul as substance that we find the most striking
doctrinal differences between the two works. In the commentary, I, 3 (pp.
248-51), in determining whether the intellectual soul is a *hoc aliquid*, he first
refutes five positions based on the difference between *quod est* and *quo est* in
the soul. The sixth position he mentions is that of Avicebron, *Fons vitae* 3,
24, that every creature is composed of matter and form, whether spiritual,
mediate, or corporeal, spiritual matter having no dimensions and not being
the subject of transmutation:

> Adhering to this opinion or position, we say and posit that in
> all spiritual substances there is spiritual matter and form,
> and thus all things are composite. And therefore one must

---

[10]   *History of Christian Philosophy in the Middle Ages*, p. 681.

[11]   *Commentarius* I, 3, *ed. cit.*, p. 247; *Scientia* I, 2, *ed. cit.*, p. 59.

say that in the intellective soul there is a composition of
matter and form, or "from which it is."[12]

This explicit hylomorphism is much attenuated in the *Expositio*, and there is
no mention of Avicebron. Peter now holds only that there is something like
matter and something like form in the intellect, although he still states without
qualification that the intellective soul is a *hoc aliquid*.[13] But in the *Scientia* he
adopts quite a different position, which was not among those discussed in the
commentary. Substance, he says, has four components: matter, form, the
composite, and essence; but the soul lacks matter. Matter receives perfection,
has form conferred upon it, undergoes, does not move itself, and is not
sufficient for any operation. But the soul bestows perfection of being and the
actuality of operation, and it therefore does not participate in the *ratio* of
matter. As John Blund had held earlier and as Anonymous Vaticanus and
Siger of Brabant would hold later, Peter says that the soul is composed of
genus and differentia. "For genus and differentia possess the same *ratio* in the
constitution of species that matter and form do in the constitution of a natural
composite subject."[14] He refers to the rational soul somewhat ambiguously as
a "substantia quodam modo completa existens,"[15] but the words "quodam
modo" would indicate that it is not quite complete in the full sense of the
word.
This is because it is truly the substantial form of the body. The body is not,
strictly speaking, considered a substance by Peter. It has no proper form of its
own as body but serves as 'suitable matter' (*materia accommodata*) to receive
the actuality (being and life) conferred by the rational soul. Peter avoids

---

[12]   " Huic autem opinioni vel positioni adherentes, dicimus et ponimus quod in
omnibus substantiis spiritualibus est materia spiritualis et forma et sic omnes composite sunt et
ideo dicendum est quod in anima intellectiva est compositio ex materia et forma vel ex quo
est." *Commentarius* I, 3, *ed. cit.*, pp. 251-52.

[13]   "Item intellectus quo intelligit homo sive anima intellectiva, non solum est
forma sed hoc aliquid; sed omne tale causatum habet duo in se, scilicet, unum materiale et
reliquum formale." *Expositio, ed. cit.*, p. 325; "Et dicendum quod sicut ens sensibile duo
habet in se, scilicet, materiam et formam, similiter et ens spirituale duo habet in se, quorum
unum est materiale et reliquum formale." *Ibid.*, pp. 328-329.

[14]   "... eamdem enim rationem obtinet genus et differentia in constitutione speciei
quam habet materia et forma in constitutione subiecti compositi naturalis." *Scientia* I, 2, *ed.
cit.*, p. 59.

[15]   *Loc. cit.*

referring to the vegetative and sensitive powers that prepared matter for the
reception of the soul as forms or souls. They are, in his view, powers that
were deposited with the sperm by the father, and when they have performed
their preliminary shaping function, they cease. They were in any case not
powers of the embryo, but of the father.[16] The rational soul, from the instant
of its simultaneous creation and infusion, performs all the vital functions. In
the commentary Peter is less than certain about the soul's being one in essence
but multiple in faculties and says he is upholding it for the sake of the
disputation.   But in the *Scientia* he adopts this position without
qualification.[17]

Peter, unlike Albert the Great, does not shrink from using the word form for
the soul. It is more appropriate than either perfection or act, both of which
may, however, be appropriately used of the soul in some of their meanings.
But like all of those who maintained that the soul was a substance as well as
being the form of the body, he had to redefine the word form. Form has two
meanings, he says. "There is the form that is only form (*forma tantum*), and
the form that is form and substance. Form that is only form has neither
operations nor passions that come forth from it without matter. But the form
that is a substance has operations and passions beyond matter as well as in
matter."[18]   The soul is the substantial form of the body, not an accidental
form, since it acts as the first regulating principle, imposing necessity on the
operations.[19]   Although the soul is a distinct substance, it is not a separate
substance but is infused into the body at the instant of its (the soul's) creation,
so that it never had a previous separate existence apart from the body. It is
joined to a material body both by existence and faculties. Peter says that it is
"a substance existing complete in a certain way," but with respect to the body
it is the formal completive principle.

> It is the form of a perfectible body, constituting it in being.
> ... Therefore the soul is called the form or perfection of

---

[16] *Commentarius* II, 11, *ed. cit.*, pp. 766-67. The fullest treatment of Peter's doctrine
of the origin of the soul is J. M. Da Cruz Pontes, *Pedro Hispano Portugalense e as
controvérsias doutrinais do século XIII. A Origem da Alma* (Coimbra, 1964). Compare the
similar views of Alexander of Hales and Roland of Cremona.

[17] *Commentarius* I, 4, *ed. cit.*, p. 263.

[18] *Commentarius*, Praeambulae, *ed. cit.*, pp. 84-85.

[19] *Commentarius* I, 3, *ed. cit.*, p. 242.

bodies because, perfecting the body as its subject, a composite of itself and the body, it constitutes a perfected thing in natural existence; and through itself and through the differentia that is founded in it, the genus is perfected.[20]

The soul is an intrinsic principle perfecting the body and is the root of all motions and operations of the body, and thus it possesses the *ratio* of form or intrinsic perfection. For through the union of its substance, it constitutes the body in specific being; through the powers applied to the parts of the body, all operations emanate from it.[21]

He very clearly states that there is in man one form only, and this form is the rational soul. "Therefore the soul alone is the perfection of the living body. ... [J]ust as matter exists in potency to the act by which it is perfected, thus the body is ordered in potency to the life that it receives from the soul, which is its perfection; and thus the body of which the soul is the perfection is a potency participating in life."[22]

So far Peter has been emphasizing the intimacy of the matter-form relation between body and soul. But, like Philip the Chancellor, he also teaches that because of the great distance between the corporeal body and the spiritual soul, media are required for their union, and form is only bestowed upon a subject that has the requisite disposition. But Peter's specification of the nature of these media is different from Philip's account and seems to have been influenced by Peter's medical training. Instead of the corporeity, simplicity and incorruptibility that Philip had employed as unifying media,

---

[20] "[F]orma enim specificum esse subiecto conferens... . Anima autem forma ac perfectio corporum dicitur eo quod corpus perficiens subiectum, ex ipsa et corpore compositum, perfectum constituit in existentia naturali et per ipsum vel per differentiam que in ipsa fundatur, genus perficitur." *Scientia* I, 2, *ed. cit.*, p. 59.

[21] "Anima vero est perfectio intrinseca, quia est principium intrinsecum corpus perficiens, et est radix motus omnis et operationis in corpore; et sic forme et perfectionis intrinsece optinet rationem. Ipsa enim per sue unionem substantie corpus in esse specifico constituit; per suas virtutes corporis partibus applicatus ab ipsa operationes emanant." *Scientia* I, 2, *ed. cit.*, p. 60.

[22] "Anima igitur est perfectio solius corporis viventis... . Sicut igitur materia existit in potentia ad actum qua perficitur, sic corpus potentia ordinatur ad vitam quam ab anima que est eius perfectio suscipit et sic corpus cuius anima est perfectio, est potentia vitam participans." *Scientia* I, 2, *ed. cit.*, p. 62.

Peter presents a scheme based on the necessity of reducing the complexion of the composite to equality.

This is done quite simply in the vegetable kingdom and in a more complex fashion in the beasts. But since the intellectual soul is separated in the highest degree from the corporeal nature, more and higher media are required to bring about an equality of complexion. The highest agency in nature is light, which shares in both the spiritual and corporeal natures, and it is this that completes the union of the rational soul with the human body. Through its agency, "the soul is united to its subject as form to matter and perfection to perfectible, because it confers being upon it and is united to it as an intrinsic mover to a mobile, and because in itself, through its faculties, it is the principle of operations."[23] However, it is not a material form. It is united to the body through its substance without continuation, contiguity, mixture, distension, or impression, so that it gives perfection to it and to each of its parts. But the *ratio* of its union is diversified in its differences according to greater dependence and coherence—less in the vegetative, greater in the sensitive, and greatest in the intellective, which seems to have no dependence, "much like the union of light and air."[24] The human soul is the highest of natural forms, and it contains all the powers that the lower forms possess. "It joins together the natures of all things, to which everything is finally ordered, so that it might deservedly be called a small world."[25]

Peter's understanding of the agent and possible intellects was, like that of many of his contemporaries, influenced by notions derived from Aristotle, Avicenna, and Augustine. He taught that the intellective soul has a twofold gaze: one toward the creator, from whom came into being everything it knows, since he is its cause. It also has a gaze toward substances similar to itself, separated from matter, and toward the body it directs and those things that are ordered to the body, which is beneath it. As a result of these two aspects, it has a twofold potency: one through which it is compared to higher things and through which it is suited to be separated and which is illuminated by the light of the soul itself. This potency is the agent intellect, and this is proper to the soul. In this it does not need the body or matter. But its other potency is the possible intellect, through which it both knows the body and

---

[23] "anima subiecto unitur ut forma materie et perfectio perfectibili secundum quod ei esse confert et ei unitur, ut motor intrinsecus mobili, secundum quod in ipso per suas virtutes est principium operationum." *Scientia* I, 5, *ed. cit.*, p. 75.

[24] *Scientia* I, 5, *ed. cit.*, p. 76.

[25] *Scientia* I, 1, *ed. cit.*, p. 55.

lower things and rules the body. Because the intellective soul is ordered for the knowledge of the creator, it is disposed to him in two ways: both immediately and *per posteriore* through its effects and operations; and thus it knows through organs, and knowledge arises in it from the potential intellect. Just as the status of the intellect is twofold, thus there is a twofold operation of the soul itself, one common, the other proper. One is called the agent intellect, the other the possible. The higher power, the agent intellect, directs the lower possible intellect by exciting and illuminating it, and by leading it to its effect both to illuminate and direct itself. In this way it depends on the body to some extent, and to this extent understanding is not completely the proper operation of the soul. But it also knows separated things, and the more removed it is from the body, the more truly it knows them. In the Madrid codex this question is treated much more fully than in the Cracow,[26] and Peter has added some remarks on the possible intellect similar to Roger Bacon's early view, denying that it, like the agent intellect, is immortal, separable, perpetual and unmixed,[27] and a purely Avicennan account of the agent intellect. In addition to the agent intellect that is proper to the individual soul, Peter says here, there is a separate angelic Intelligence, without whose cooperation abstractive cognition would not be possible. This active substance is the lowest of the separated Intelligences, and, being intimately present to the intellective soul, reveals the intelligibles to it by its illumination, so that its relation to the intelligible forms is the same as that of light to colors.[28] On this passage, Gilson has commented: "Peter does not reduce the separate Intelligence to the God of St. Augustine; he states its existence and its illuminating function just as they are described in Avicenna. To our personal knowledge this is the only known case of straight Latin Avicennism."[29]

It is difficult to categorize Peter's doctrine of the soul. His view of matter is clearly that which underlay theories of the plurality of forms. But he also explicitly upholds the unity of substantial form. His explanation of the agent

---

[26] See *Commentarius* I, 6, *ed. cit.*, pp. 294-95 and *Scientia* X, 5-7. *ed. cit.*, pp. 429-50.

[27] *Scientia* X, 6, *ed. cit.*, p. 443. This is also the view expressed in the *Expositio*, *ed. cit.*, p. 324: "... intellectus agens solum et immortale et perpetuum ...; et per hoc quod dicit solum intendit excludere intellectum possibilem. ... *Non reminiscimur autem*. Hic per signum dicit quod intellectus possibilis non est separabilis et perpetuus ... quia post separationem animae a corpore non reminiscitur anima."

[28] *Scientia* X, 7, *ed. cit.*, p. 446.

[29] *History of Christian Philosophy in the Middle Ages*, p. 322.

and possible intellects grafts Augustinian and Avicennan teaching onto Aristotle, but in a different fashion than any of his contemporaries. Most of them held that the agent intellect was a faculty of the individual soul; some equated it with God; some admitted a higher illuminating agency in addition to the agent intellect proper to the soul but always equated it with God, the 'uncreated light' of Augustine. Peter holds that there is an agent intellect proper to the soul, but that there is also an angelic Intelligence that makes intellective cognition possible.

Peter made a truly ingenious attempt to explain how a substance could at the same time be a form. Although the soul is a substance, it lacks matter, and it is created specifically to be the form of the human body, which is the matter of the soul-body composite. But it is still the body, not prime matter, that is the subject of the soul. Matter must be suitable to the form that it receives, and so only a pre-formed body fulfills this requirement. He avoids the problem of what happens to the vegetative and sensitive powers upon the arrival of the rational soul, as well as the question of the corruption of previous forms by something other than their contraries, by denying that the vegetative and sensitive powers are souls or forms, or that they belong to the embryo.

Peter and Albert the Great both suggested that the soul is not strictly speaking a complete *hoc aliquid* (although both men on occasion also used this term for it), Peter in emphasizing its status as a form, Albert by emphasizing its natural affinity for the body. William of Baglione, Bonaventure, and Aquinas would develop this suggestion further. The doctrine that the soul must be joined to the body by a medium came from the *De differentia spiritus et animae* by Costa ben Luca. It seems to have been adopted by Peter in order to avoid making the soul a material form, while still allowing for it to be the form of matter. He goes to considerable lengths to show that it is utterly unmixed with the body—there is no continuity, contiguity, mixture, distension, etc.—but is still the perfection of the body and the source of all its operations.

Although it is undeniable that Peter's doctrine of the intellect was heavily influenced by Avicenna and so contrary to Aquinas's teaching, on the question of the relation of soul and body he came closer than anyone else to Aquinas's solution. He suggests that the soul is not strictly speaking a *hoc aliquid*; he speaks of it as giving being as well as life to the body and being its substantial form; and he does not consider the body to be a substance, with its own sub-rational or corporeal form, prior to the infusion of the rational soul. This puts him at least as much at odds with the "Augustinians" as he was with Aquinas on the doctrine of the intellect.

ROGER BACON. I am always reluctant to include Roger Bacon in any discussion of medieval thought or to use his testimony as evidence for scholastic events, since his thought processes were often perversely idiosyncratic, and his tendency to exaggerate sometimes borders on outright lying. But he was an extremely intelligent, if somewhat erratic, thinker, and he was among the first generation of those artists who lectured on the corpus of Aristotle's natural philosophy. His mind was quintessentially eclectic and creative rather than critical and analytical, and although he considered that he was interpreting the text of the Stagyrite correctly (although he frequently complained about the faulty translations of his works), it is clear to modern readers that, like that of his contemporaries, his understanding of Aristotle was distorted by the influence of Avicenna, Avicebron, Boethius, Alfarabi, and Gundissalinus, and the misattribution of several important works, especially the *Liber de causis* and *Secretum secretorum*. Bacon's tendency to synchretize the views of his authorities (although he was sometimes conscious of inconsistencies) often prevented his getting the doctrine of any of them right.

Fortunately, however, there exists a first rate learned and judicious guide to Bacon's doctrine of the soul by Theodore Crowley,[30] to which I am much indebted for what follows. One of Crowley's main contentions (and I have found no reason to contradict it) is that Bacon considered himself an Aristotelian and would always prefer what he thought to be Aristotle's view to any conflicting position of his other authorities. He was in no sense an Augustinian, although he adopted several positions, such as universal hylomorphism and the plurality of forms, which would be key components of the 'Neo-Augustinianism' of the later part of the thirteenth century. And although he lectured in the arts faculty, he was always keen to avoid any position that might be considered heretical. He had no inclination to push philosophy into positions where it might seem to contradict faith.

Since we are interested in Bacon's thought as a testimony to the character of studies of the rational soul during the 1240s, the period of his Parisian arts regency (1241-46), we shall confine ourselves to this period and not explore later developments in his thought.

One of the earliest basic positions taken by Bacon was that of universal hylomorphism. He considered it to be authentically Aristotelian and never abandoned it. This posed for him the same problem it did for many of his

---

[30] Theodore Crowley, *Roger Bacon. The Problem of the Soul in his Philosophical Commentaries* (Louvain/Dublin, 1950). Also very helpful is the subsequent but more limited study of Efrem Bettoni, "Origine e struttura dell'anima unana secondo Bacone," *Rivista di filosofia neo-scolastica* 61 (1969), 185-201, which emphasizes the incoherence of Bacon's doctrine of the soul.

contemporaries and successors: if the soul is an independent substance—a *hoc aliquid*—composed of matter and form, how could it function as the substantial form of the human body and constitute a single substance with it?

Bacon considered the human soul to have a dual origin. The vegetative and sensitive souls are educed from the potency of matter and are *ex traduce*. Unlike the sensitive soul in beasts, that in man is incomplete. It and the vegetative soul are dispositive forms that prepare the body to receive the rational soul, which is 'from without,' a special creation by God. The rational completes and unites with the lower souls, which however remain operational, to constitute a unified but compound soul. Bacon contemptuously denies the opinion of those who hold that there is a double vegetative and sensitive soul, one *ex traduce*, one the created rational soul, which duplicates the vegetative and sensitive functions. He therefore holds to the unity of substantial form and considers the rational soul to be the substantial form of the body, but it is a compound, not a simple, form. The soul thus constituted is both corruptible and incorruptible. The rational soul alone is created in the image of God and is a *hoc aliquid*; the vegetative and sensitive are natural forms and thus corruptible. He tells us in his commentary on the *De causis*[31] that he had proved in his *De anima* (which has not survived) that the vegetative, sensitive, and rational principles are three simple essences (or two distinct and diverse essences), which unite to form one essence, no longer simple, but compound. Just as matter and form—also two simple essences—unite to form one substance, so too the eternal and the generated can be united in one subject.[32] This division of the soul into simple constitutive parts, or essences, would be brilliantly adapted and made more rigorous by Anonymous Vaticanus.

Bacon had a great deal of difficulty in understanding what Aristotle meant by the possible and agent intellects. In fact, as we have noted previously, it is not at all clear what Aristotle meant in *De anima* 3, 5, but Bacon's changes of mind seem to have been occasioned by what his subsidiary authorities taught and his attitude toward them rather than by an analysis of Aristotle's text. Early in his career, Bacon taught that the agent intellect was a part of the individual human soul, elevated to the contemplation of higher things, and it was this agent intellect that remained after the death of the body. The possible

---

[31] *Opera hactenus inedita fratris Rogeri Baconis* (Oxford, 1905-1941), XII, p. 158.

[32] Crowley, *Roger Bacon*, pp. 137-41 attempts to reconstruct Bacon's reasoning as it would have appeared in the lost *De anima*.

intellect, although it was the same in substance, was worn out by the effort involved in pursuing understanding, and it perished with the body.[33]

Later, he moved in the direction of Avicenna's position in holding that the agent intellect is a separate substance and not part of the soul. He did not however equate it with the Giver of Forms, as Avicenna had done, and in fact he says very little about it, only that it is not a part of the soul and that its function is to make actually intelligible the potentially intelligible species supplied by the senses. Like all his contemporaries, however, he attributes to Averroes the position that the agent intellect was a part of the soul. But since he now taught that the agent intellect was not part of the soul, Bacon had to revise his view of the possible intellect, to which he had previously denied immortality, and attribute to it many of the characteristics he had previously assigned to the agent intellect.[34] Bacon's final position was to identify the agent intellect with God. This was his view when he composed the three *Opera*.

If Bacon's treatise *De anima* had survived, his teaching on the human soul might appear more coherent than it does from the collection and comparison of chance remarks scattered among his Aristotelian commentaries. But I doubt it. Like all his contemporaries, he had the all but impossible task of reconciling the view of the soul as a substance in its own right with the view that it was the substantial form of the body. He had an inadequate comprehension of Aristotle's doctrine, even allowing for ambiguities and

---

[33] "Alius est intellectus creatus materie transmutabili coniunctus, scilicet corpori, et hic est duplex: quidam est agens, scilicet una pars intellectus elevata ad superiora contemplandum, et hec vocatur intellectus agens, et hec non intelligit per administrationem sensuum, set per exempla sibi innata, confusa tamen; et quantum ad hanc partem non suscipit intellectus lassitudinem, langorem in intelligendo, et hic est intellectus agens <qui> remanet in anima quando a corpore separata est. Alter est intellectus possibilis, scilicet altera pars intellectus vel rationis quando ratio se inclinat ad inferiora, et hic intelligit per administrationem sensuum, de quo dicitur 'nichil est in intellectu quin prius fuerit in sensu;' de quo dicitur, 'omne nostrum intelligere est cum continui et tempore.' Et hic lassitudinem et fatigationem, langorem suscipit in consecutione intelligendi; set non agens quamvis idem sint in substantia quia intelligere agentis non est mensuratum a tempore." *Comm. in Metaph.* XI, *Opera hactenus inedita* VII, p. 110.

[34] "RESPONDEO quod duplex est hic agens, universale et particulare; a parte intellectus speculativi agens universale et particulare exigitur, scilicet intellectus agens; respectu intellectus practici non ponitur Aristotele agens particulare set universale, quia philosophice loquendo in omnibus operationibus naturalibus et spiritualibus causa particularis non potest agere sine influentia cause prime et universalis. Unde intellectus agens est illud agens particulare quod exigitur ad operationem intellectus speculativi, qui quidem intellectus agens secundum Commentatorem est pars anime, secundum Alpharabium et secundum Aristotelem et Avicennam est aliquid aliud." *Quaestiones super nonum librum Metaphysice. Opera hactenus inedita* X, pp. 298-99.

inconsistencies in the Stagyrite's text, and his radical hylomorphism vitiated any attempt to depict the soul as the substantial form of the body. He even had trouble deciding what the soul was. Although in this life it was a unity compounded of three distinct essences, two of them corruptible and one incorruptible, after separation from the body only the rational part remained, and Roger was not certain whether this included the agent intellect or not.

None of the elements of Bacon's teaching on the soul was original with him, but he seems to have put them together in a uniquely inconvenient way. His doctrine confirms what we have seen in writers from the beginning of the century onward, and makes it clear that by the 1240s, scholastic thinkers could not agree on what the human soul was.

ANONYMOUS ADMONTENSIS. An anonymous master of the Parisian Arts faculty composed a set of questions on *De anima*,[35] dating from between 1250 and 1260. Although it lacks the sophistication of the works of the next two decades, it is a major advance over the preceding treatments of the soul by those artists we know of.

The author uses two translations of *De anima*, designating the *translatio vetus* as *translatio nostra* and the *nova translatio* as *translatio alia*. In addition to Aristotle's works, he cites Avicenna's *De anima*, Averroes's *De substantia orbis* and commentaries on *De anima*, *Physica*, and *Metaphysica*, Porphyry's *Isagoge*, Boethius's commentary on the *Categories*, and his *Liber divisionum* and *De consolatione philosophi*ae, Avicebron's *Fons vitae*, and (through Averroes) Theophrastus, Alexander of Aphrodisias, Themistius, Alfarabi, Avempace, and Abubacer. There are no citations of thirteenth-century authors by name, although there is one citation of the *De intelligentiis* of Adam Pulchrae Mulieris or Adam de Puteorum Villa.[36]

---

[35] Joachim Vennebusch, ed., *Ein Anonymer Aristoteleskommentar des XIII. Jahrhunderts. Questiones in Tres Libros De Anima (Admont, Stiftsbibliothek, cod. lat. 367)* (Paderborn, 1963).

[36] This work, also known as *Memoriale rerum difficilium*, was attributed to Witelo by Clemens Baeumker, *Witelo, Ein Philosoph und Naturforscher des XIII. Jahrhunderts*, BGPM III.2 (Münster i.W., 1908). But Aleksander Birkenmajer shows conclusively that the author cannot have been Witelo and is named as Adam Puchre Mulieris in two manuscripts, and that the work was written around 1230 and was already well known by 1250, in "Witelo est-il l'auteur de l'opuscule 'De intelligentiis'?" *Studia Copernicana* 4 (1972), 259-335, reprinted in French translation from the original, which appeared in the *Bulletin de l'Académie Polonaise des Sciences de Cracovie*, 1918 (published 1920). The name of the author is more probably Adam de Puteorum Villa; cf. P. Glorieux, "Maître Adam," *Recherches de théologie ancienne et médiévale (1967), 262-70*.

The author dismisses the hylomorphic composition of the soul. It does not however lack all composition, for there must be in it something corresponding to matter so that it can undergo the passions of the soul and be receptive of forms; and there must be something corresponding to form so that it may be the act of the body.[37] He identifies *quo est* and *quod est* as that from which the soul is composed.[38]

· He also attempts a solution to the problem of the soul's apparent double origin, examining in detail and rejecting the position that there are three souls in man (since this would result in three composites, not one), as well as that which holds that there is one soul, which corrupts the earlier vegetative and sensitive souls of the embryo when it is infused into the body, because a thing is corrupted only by its contrary. In our author's solution, he first investigates the relation of the vegetative and sensitive souls in beings that lack the intellective. In such beings, the vegetative soul begins the organization of the embryo but is incomplete in itself and engenders the complementary and perfecting sensitive soul. The sensitive soul then behaves as the perfection of the substance. But, he says, it is otherwise in things possessing an intellective soul. "The sensitive soul does not behave toward the intellective as an incomplete form, from which the intellective is generated through the eduction of a complement; but the intellective is a certain substance completely separate, which arrives to the sensitive as form to matter,"[39] since the sensitive soul alone is in receptive potency to the intellective. The sensitive soul, educed from the potency of matter, can never be of the same essence as the intellective soul, which comes from without. Nevertheless from the sensitive and intellective is made one soul. The sensitive behaves as the material constituent, and the intellective as the formal, constituting one soul, which behaves as one act and one perfection with respect to an organic body, "so that the soul of man, from which he has one being in act, is not called only that which is induced from without, but the whole aggregate educed by the intellective and sensitive souls from preexisting matter. ... And thus from these [two] is made one soul *per essentiam*, although such things do not become one

---

[37] Q. 31, sol., III, *ed. cit.*, pp. 176-77 and Q. 69, ad 2, *ed. cit.*, p. 205.

[38] Q. 69, 2, contra, *ed. cit.*, p. 304.

[39] "... de intellectiva respectu vegetative et sensitive alia est ratio. cum enim non se habeat sensitiva ad intellectivam tanquam forma incompleta, ex qua per educcionem complementi habeat generari intellectiva, sed intellectiva est substancia quedam penitus separata que advenit sensitive sicut forma materie." Q. 31, sol. III, *ed. cit.*, p. 176.

essence; from this man is said to be one in his being, and not several according to several acts."[40]

Man's substantial form then is single but not simple. The sensitive soul, which had organized the embryo to the point that the intellective could be infused, behaves as matter toward the intellective, which acts as form toward it. The composite is of dual origin, and the marks of dual origin remain in the composite soul. The intellective subsumes the sensitive not as superior form to inferior form, but as formal principle to material principle. It is nevertheless one soul, and in its material part (the sensitive) it had organized the matter of the embryo; then, having a natural potency to be perfected by the intellective, it performs the function of matter in the composite. This is a considerable improvement over previous attempts to describe the compound human soul. It attacks the same points of Albert the Great and Aquinas that Siger would object to to the end: that the vegetative and sensitive souls are of the same substance as the intellective; and that preceding forms could be corrupted by something other than their contraries.

Our author considers the possible and agent intellects to be part of the individual human soul, and, like Philip the Chancellor, Roger Bacon, Albert the Great, Peter of Spain, and Adam of Buckfield, he attributes this interpretation to Averroes.[41]

ANONYMOUS VATICANUS. Joachim Vennebusch has also discovered a series of questions on books 1 and 2 of De anima in MS Vat. lat. 869, folios 200r-210v,[42] which marks a further development along the same lines as the doctrine of Anonymous Admontensis. He does not categorically state that the author of both works is the same man, but he strongly suggests this and presents quite a good case for it. But since their identity cannot be definitely asserted at the present time, it seems better, provisionally at least, to consider them as two distinct authors. Vennebusch has printed questions 25, 26, 27, and 28, all dealing with the problem of the unity of the soul, and he has provided a most helpful study of the doctrine of the work. He suggests that

---

[40] "...ita quod anima hominis a [qua] habet esse unum in actu, non dicitur solum illud quod ab extrinseco est inductum, sed totum aggregatum ab intellectu [et sensitiva exducta] ex precedenti materia ... et sic ex istis fit una anima per essenciam, licet ista non fiant una essencia; a qua dicitur homo esse unus in entibus et non plura secundum actus plures." Q. 31, sol., III, ed. cit., pp. 176-77.

[41] Q. 67, III, ed. cit., p. 297.

[42] Joachim Vennebusch, "Die Einheit der Seele nach einem anonymen Aristoteleskommentar aus der Zeit des Thomas von Aquin und Siger von Brabant (Vat. lat. 869, ff. 200r-210v)," Recherches de théologie ancienne et médiévale 33 (1966), 37-80.

the author may be connected with the Oxford circle of masters, and this seems likely enough but beyond proof. To his evidence for this identification, I would add that Anonymous Vaticanus is much concerned with formal logic, a characteristic of the English masters, and that he devotes a total of nine questions to the investigation of the nature of light, a subject that had preoccupied Grosseteste, Bacon, Kilwardby, pseudo-Grosseteste, and Pecham.

As to the date of the work, Vennebusch says only that both it and Anonymous Admontensis seem to come out of the milieu of the decade 1250-1260, and that Anonymous Vaticanus is the later. If these questions are indeed as late as 1260, it is surprising that the author seems ignorant of the doctrine of the uniqueness of the intellect, which by that time was widely recognized as being Averroes's interpretation of Aristotle. While it is true that these questions cover only books 1 and 2 of *De anima*, other masters managed to work this topic into their questions on the first two books if they wished to discuss it. If Vennebusch is correct, the author knew of Aquinas's doctrine of the progressive succession of substantial forms during the development of the embryo and their annihilation at the infusion of the rational soul. I think however that Vennebusch is wrong in believing that these questions show the influence of Siger of Brabant or the so-called 'Latin Averroists.' They probably do however give us a good notion of the teaching on the soul in the arts faculty while Siger was an undergraduate.

This is a subtle and rigorous work, attempting to find a middle ground between what the master considers two untenable positions: that there are three distinct souls in man; and that there is one simple substance of the soul, possessing three powers. But he also spends some time disproving the 'two-essence doctrine' that was espoused by Anonymous Admontensis. Whether this represents a development of his own thought, or simply includes an earlier unsatisfactory attempt by someone else to do the same thing, is not clear.

Some comments on vocabulary are necessary before we look at the work itself. This master uses the terms 'essentia simplex' and 'substantia' synonymously. Although Roger Bacon had also resorted to an idiosyncratic use of 'essentia' in his explanation of the composition of the soul, he had considered essences to be the constituents of a substance, and he had explained his use of the term by an analogy to matter and form, two essences that make one substance. Anonymous Vaticanus uses the term to mean a simple substance, such as that of the vegetative, sensitive, or intellectual soul taken in itself, over against a 'substantia aggregata,' or substance made up of simple substances, such as the human soul considered in its entirety. He also made a distinction between something being of one essence and being one through essence (*una essentia* as against *unum per essentiam*).

Question 25 is devoted to determining in what way the vegetative, sensitive, and intellective are divisions of the soul. This is a key element in his doctrine, by which he attempts both to preserve the unity of a substance and to allow the simple intellectual essence to have a power not involved in matter. He begins by making a distinction between the substance of the soul and the powers of the soul. In considering the substance of the soul, there is a division of genus into species, and these three simple substances (vegetative, sensitive, and intellective) are diverse in species. In the sensitive soul of a beast, the sensitive is the formal element, the vegetative the material, and the aggregate of these two make one perfection. Similarly, the intellective, according as it names the substance of the soul that is the perfection of man, names something made up of three substances of the soul, which when aggregated make one perfection, although it is composed of three essences or substances of the soul. And the word 'intellective' names not only the intellective substance but something composed of three substances of the soul. The intellective is the formal constituent, the other two the material. In each level of soul, the part that is the perfection gives its name to the whole soul. There is thus a division of genus into species, because the thing that is divided has the mode of genus and the thing that divides it the mode of species. The soul, insofar as it is one act or perfection, is in the genus as a principle or differentia.

But when these terms, vegetative, sensitive, and intellective, name powers rather than substances, there is a division of the virtual whole into its powers. This division must be reduced to a division of the genus, because it is by the mediation of the division of the genus into species that the division of the virtual whole into powers is made, because there is a power proper to each substance of the soul: there is a power that is the perfection of a plant, a power that is the perfection of a beast, and a power of the substance of the soul that is the perfection of man. And for this reason, that which is divided into the substance of the soul, or species, must be divided into powers of these substances as the virtual whole into its powers, by the mediation of species.

A third division is that of the integrated whole according as it comes to be in the parts that integrate the essence and name simple essences of the soul and are in numerically one and the same thing.

And a fourth, similar to the reduction of the powers of the virtual whole to the division of the genus into species, is a division of the virtual whole into powers. This division must be reduced to the division of the integrated whole, because it comes about through its mediation.

Following upon this, question 26 asks whether there is one substance of the soul in the same man, or whether they are diverse. The author begins by stating three current answers to the question. The first is that there are three distinct souls in one man, and he says that this is obviously false, since there can only be one perfection of one animated body, and one ought not to waste any time disproving it.

He spends much more time on the second: that of "Augustine and his followers," that in man there is one simple essence of the soul, which is the perfection of man, and that there are three powers of this soul, vegetative, sensitive, and intellective, which the soul brings with it when it arrives and takes with it when it is separated. If one objects to those holding this position, he says, that it is necessary that the body be prepared, arranged, and organized before the reception of the intellective soul, that this can only be done by vegetative and sensitive powers, that these must come from that which generates the body (i.e., the parents), and that, since the intellective comes with its own vegetative and sensitive powers, there would be in the same thing two vegetative and two sensitive powers; they respond by saying than when the intellective soul arrives (bearing all three powers), the previous vegetative and sensitive powers are corrupted. Although we have noted this explanation in the teaching of William of Auvergne, John of La Rochelle, Peter of Spain, and Albert the Great, Vennebusch thinks that the reference here is to Aquinas, since only he used the word 'corrumpuntur' to designate the disappearance of the previous powers.[43] In this case, Anonymous Vaticanus would be ranging Aquinas among the Augustinians.

In any case, he says, this opinion cannot stand, for four reasons. First, because that which is in another can only be corrupted through the corruption of the subject in which it is, or through the arrival of a contrary. But with the arrival of the intellective soul, the already-disposed body is not corrupted but is rather completed. Second, the intellective soul is not the contrary of the dispositive powers, nor does it bring with it anything contrary to them, and so they cannot be corrupted. Third, that they would not be educed from the potency of matter and should yet become an act or perfection is contrary to Aristotle; and thus, just as in the previous case, there will be two vegetative and two sensitive powers in the same man. And fourth, the vegetative and sensitive are corrupted either while they move the body or according as they are its act and perfection. Not the first way, because Aristotle says that that which moves something becomes the act and perfection of the thing that it moves. And not according as they are already the form or perfection, because a substantial form can only be corrupted through the corruption of the composite of which it is the form. But the vegetative and sensitive are substantial forms and perfections, although they are not completive, and thus they cannot be corrupted except through the corruption of the composite. But the composite is not corrupted; indeed it is rather preserved. Therefore, neither the vegetative nor the sensitive can be corrupted by the arrival of the intellective. They could not be corrupted in an instant, because the agent

---

[43] Vennebusch, *art. cit*, p.50.

bringing about the corruption must precede what it corrupts. It could not be corrupted before the arrival of the intellective, because then the body would be without a soul for a period of time. And they could not be corrupted after the arrival of the intellective, because then for some time there would be two vegetative and sensitive souls. "It is clear therefore that one ought not to hold this position, and is is not in accord with the philosophy of Aristotle; therefore it should be denied." [44]

This is by far the most thoroughgoing attack on the single substance of the soul theory that we have encountered, more so than even that of Siger will be.

The third position against which Anonymous Vaticanus argues is essentially the one put forth by Anonymous Admontensis: that the vegetative and sensitive are the same in essence because the vegetative becomes the sensitive, since they are related as complete and incomplete. And thus the soul of man is composed of two essences, the sensitive and the intellective. But this position, he says, is nothing, because the vegetative and sensitive are not related as incomplete and complete, but as incomplete and complement. An incomplete thing never becomes a complement but comes to be under a complement, just as the form of a genus is related to a differentia. And the form of a genus does not become the form of the differentia but comes to be under the differentia. Therefore, the incomplete and the complement always differ in essence; the form of the genus becomes the form of the species plus the assumed differentia or complement. And the form of the species is called complete with respect to the form of the genus and differentia. The form of the genus does not differ from the species in essence, because it is included and contained in it. They are not completely the same, because they are related as essential parts to the whole, genus and differentia with respect to species. Just as the form of the species is something composed of an incomplete and a complement, thus the form of man is composed of an incomplete and a complement. The prior form is something incomplete with respect to the form that succeeds it, and the form that succeeds it is related as a complement, not as a completed thing. Thus it is incorrect to say that the vegetative becomes the sensitive; rather, it comes to be under the sensitive.

In his criticism of the third theory, Anonymous Vaticanus has in fact begun the exposition of his own solution, to which he now turns: that the vegetative, sensitive, and intellective, according as they are in the same thing, as in man,

---

[44]   "patet igitur, quod positione tali non est adherendum, neque est secundum philosophiam Aristotelis." *Ed. cit.*, p. 77, ll. 234-35. In his reply to the fifth argument, he says: "Quod arguitur per Augustinum, quod anima hominis est essentia intellective simplex, cuius sunt ille tres potentie,— dicendum quod non est illi adherendum secundum Aristotelis sententiam, et ideo negandum est." *Ed. cit.*, p. 79, ll. 309-12.

are not one essence, but they make one thing *per essentiam*. And he insists that his is a proper distinction to make.[45] They are in man truly distinct essences, so that one is not the other. Nevertheless, those three essences thus aggregated make one perfection and one soul. The two prior essences (vegetative and sensitive) are material or potential with repect to the intellective, which is the complement. And from these, just as from potential and formal, or just as from an incomplete and a complement, is made one soul, as one perfection. They are not one essence but make one thing *secundum essentiam*. And "thus the intellective, according as it names the substance of the soul in man, names something made up of three essences. ... Since the soul of man not only names the complement, or incomplete, but something made up of all three essences, of which one behaves in the role of a complement, it is not illogical that form should name such an aggregate or composite, because specific form collects in itself the anterior forms preceding in its genus."[46]

The question of the soul's relation to the body, in this view, is clearly answered: the aggregated soul is the substantial form of a human being, the body is the matter. Each of the aggregated soul's essential parts (simple substances) has its own distinct powers. Although the master does not expressly say so, this would make it possible for the intellective part to possess the power of understanding without itself being involved in matter, although the other two parts of the aggregated soul would be so involved. This avoids the dualism that had plagued so much of Latin thought on the soul-body relationship, and allows the soul to be the substantial form of man while leaving the intellectual faculty unmixed with matter. This is a very impressive solution. Its only serious shortcoming, as far as I can see (and this from a theological point of view), is the same one we have noted in the thought of Philip the Chancellor and Roger Bacon, that the resurrected body will have only a rational soul, its vegetative and sensitive constituents having perished with the death of the body.

This work represents a type of Aristotelian exegesis that goes back to John Blund at the beginning of the century, with Blund's attempt to explain the composition of the soul as being of logical relations—genus, differentia,

---

[45] "Et refert sic dicere aut sic." *Ed. cit.*, p. 78, ll. 268-69.

[46] "sic intellectivum prout nominat substantiam anime in homine, nominat quid aggregatum ex tribus essentiis. ... cum anima hominis non solum nominat complementum sive incompletum, set quid aggregatum ex omnibus essentiis quarum una se habet in ratione complementi, nec illud est inconveniens quod forma talem aggregationem sive compositum nominet; quia forma specifica aggregat in se formas anteriores in suo genere precedentes." *Ed. cit.*, p. 78, ll. 279-86.

species. It was continued by Philip the Chancellor, who apparently claimed that one substance (soul) could behave as matter in a further composite. Both these ideas were included in the work of Peter of Spain. Roger Bacon added the notion that the single substance of the soul could be composed of three essences—vegetative, sensitive, intellective—which he envisioned as the constituents of a substance, analogous to matter and form, rather than as three 'simple substances' as opposed to a composite or aggregated substance made up of three simple substances. The notion of a composed but unified form seems also to have originated with Philip the Chancellor, and it was picked up by a succession of masters, both theologians and artists, after him. This type of pluralism should not, I believe, be confused with the later doctrine of plurality of substantial form. The former taught the unity of form while claiming that a form need not be simple, while the latter, in the words of Matthew of Aquasparta, taught that "it is necessary to posit in one and the same thing a plurality of forms, not through aggregation and accumulation, but through a certain complexion and order;" although those who maintained the latter position were surely much indebted to it. William of Baglione and John Pecham both draw extensively from this tradition, and Matthew of Aquasparta's preferred philosophical position was very close to that put forth by Anonymous Vaticanus. Siger would pick and choose elements from the tradition in his own doctrine of the soul and even try to use portions of it to accommodate the new understanding of Averroes that the intellect was unique for the human race. This is therefore a work of major importance in the development of Latin thought on the soul in the thirteenth century.

B. THEOLOGIANS

JOHN OF LA ROCHELLE. John of La Rochelle was one of the better known disciples of Alexander of Hales, although he was also greatly influenced by Philip the Chancellor. His *Summa de anima*,[47] although it was written around 1240, is a curiously old-fashioned work, only slightly touched by Aristotle's *libri naturales*. His authorities for the most part are old-fashioned: Hilary of Poitiers, Boethius, Jerome, ps.-Augustine *De fide ad Petrum*, *De ecclesiasticis dogmatibus*, and *De spiritu et anima* (he attributes all of these to Augustine), and Calcidius. He occasionally mentions more up-to-date authorities: Lombard's *Sentences*, Avicenna, Gregory of Nazianzanus, and John of Damascus, and he alludes to several Aristotelian works (*De animalibus* and

---

[47] Teofilo Domenichelli, ed., *La* Summa de anima *di frate Giovanni della Roch-elle* (Prato, 1882).

De anima), but it is not clear that he had read them. The questions he asks are more typical of the twelfth century than of his own time, but the Aristotelian notion of the soul as form has introduced considerable confusion into his teaching.

Although John vigorously asserts the unity of the human being ("From the soul and body is made one thing according to substance, which is man"[48]), his insistence on the substantial reality of the soul makes it difficult to see how this unity is achieved. Still, in arguing for its substantiality, he does not accept the doctrine of its matter-form composition, a doctrine he refutes by a quotation from Augustine.

The soul, like all created beings, is composite, but its composition, as Boethius teaches in *De trinitate*, is of *quod est* and *quo est*. There are two kinds of composition, he says, one of matter and form, the other of *quod est* and *quo est*. The latter is created from nothing, while the former is made from something already existing.[49] And in addition to being a substance in its own right, it is the form, act, or perfection of the body. Unlike the forms of inanimate objects, which depend completely on their matter and do not rule it; or the forms of shrubs and beasts, which depend on their matter and rule it and operate only through it; the rational soul rules its matter, which depends on it, but whose principal operation is not in or through matter. Its essence does not depend on the body and, although it is a form, it is nevertheless separable from the body.[50] Superficially, this has much in common with Aquinas's doctrine. as we shall see in chapter 5. John has seen that the soul must be the form of the body in order to safeguard human unity, and that it cannot thereby be a material form, in order to safeguard its immortality. Its principal power, understanding, must flow from its essence and not be involved in matter. But John merely states the *desiderata* without constructing an explanation of how the soul could be what it must be, as Aquinas did.

In discussing whether the soul is united to the body by a medium or not, John follows Alexander of Hales in wanting to have it both ways. He says that it is united to the body in two ways: as form or perfection, and in this way it is united without a medium; and as a craftsman to his tool or mover to thing moved, and in this way it is united to it through the medium of its powers. And so John holds simultaneously two incompatible views of the relation of the soul to the body: that as form it is united immediately to the body; and as a

---

[48] *Summa de anima* Pars I, 38, *ed. cit.*, p. 170.

[49] *Summa de anima*, I, 13, *ed. cit.*, p. 120.

[50] *Summa de anima* I, 35 and I, 38, *ed. cit.*, pp. 162 and 171-72.

substance in its own right it is united to it only through the operation of its powers.

John is very clear that the soul is simple, that there is only one for each person, and that all the powers of life are performed by this one soul. He opposes those who, on the ground that the first two are perishable and the last immortal, hold that there are three substances (vegetative, sensitive, and intellective) in man, by saying that the vegetative and sensitive powers in man are not corrupted in man with respect to their essence or power, but only according to their act.[51]   But somewhat inconsistently he also explicitly maintains the doctrine of double vegetative and sensitive souls, which we found to be implicit in the teaching of Alexander of Hales. "There is a double vegetative soul," he says, "one dispositive, the other perfective. The one is a form in becoming, the other in being. The first arises and is passed along in the body; the second is infused with the rational soul. The first passes away when the body has been formed and organized, but the second remains placed in the being of the rational soul. And the same can be said about the sensitive."[52]   In another question he says much the same thing, but with an important difference, concerning the sensitive soul: "The first comes into being with the body; the second is infused with the infusion of the rational soul itself. The first does not merit the name soul, since it only disposes rather than perfects; the second does merit [the name soul]. Similarly the first perishes with the body, but the second does not, because it is founded in a substance of an immortal nature."[53]   In the first of these, he says that the prior soul perishes when it has completed its task of forming and organizing the body, but in the second he explicitly says that it perished only when the body does.

But in John's thought, the subject of the soul is the body, not prime matter. It is not the soul that makes the body be a body, but rather which makes a body

---

[51]   *Summa de anima* I, 24, *ed. cit.*, p. 136.

[52]   "Duplex est vegetative, scilicet disponens et vegetativa perficiens. Una est forma in fieri, alia in esse; prima autem traducitur et seminatur cum corpore; secunda vero infunditur cum rationali animae. Prima ergo transit completo fieri, idest formato corpore et organizato; secunda autem manet posita in esse rationalis animae. Et similiter etiam distinguendum est de sensibili." *Summa de anima* I, q. 24, *ed. cit.*, p. 137.

[53]   "Prima seminatur cum corpore; secunda infunditur cum ipsi infusione animae rationalis; prima non meretur nomen animae, cum sit disponens et non perficiens; secunda meretur.   Similiter prima interiit cum corpore; secunda non, quia fundatur in substantia immortalis naturae." *Summa de anima* I, q. 43, *ed. cit.*, p. 186.   See the discussion in Zavalloni, *op. cit.*, pp. 402-03.

be a human being. Following Philip the Chancellor, he says that the human body is the most 'composed' thing among bodies and the most perfect thing constituted of the elements, and the rational soul is the most perfect perfection among all natural forms. Therefore, the human body will be unible to the rational soul. But since the body must be organic, it is not possible that it be simple or of one nature.[54]

John's notion of what the agent intellect might be is vague and confused. He casts his net wide, asking "whether the agent intellect is separate from the substance of the soul, or is a differentia of the soul, and, if it is separated, whether it is a created Intelligence (which is an angel) or uncreated (which is God)."[55] After a rather diffuse consideration of the possibilities, he concludes that for our knowledge of things higher than the soul, the agent intellect is God. For our knowledge of things on the same level as the soul, the agent intellect is angelic revelation or instruction. And for our knowledge of things that lie within the soul or below it, the agent intellect is a light innate in the soul, the *vis animae suprema*.

Although there is much that it inept in John's *De anima*, the work is interesting in being a Franciscan treatise that denies the hylomorphic composition of the soul, maintains the unity of the soul, and identifies the rational soul as the form of a human being.

ALBERT THE GREAT. The mind of Albert the Great was of a higher order than those of most of the writers we have hitherto investigated. His knowledge of Aristotle was broader and his understanding of the Stagyrite's thought was more penetrating. But he remained an eclectic, although a perceptive and discriminating one. Although Albert is most often characterized as the foremost champion of the New Aristotle in the schools, the powerful influence of Augustine and Avicenna on his thought must not be overlooked. This influence is particularly evident in his concept of the soul, which was penetrating and extensive. Throughout his career he devoted all, or a significant portion, of six works to the subject: *De intellectu et intelligibili, De natura et origine animae, De anima, De quindecim problematibus, De unitate intellectus contra Averroem,* and part 2 of the *Summa de creaturis.* Except in his *De unitate intellectus contra Averroem,* his works on the soul are written from the standpoint of a theologian, although they contain a large amount of natural philosophy.

---

[54] *Summa de anima* I, 36, *ed. cit.,* pp. 163-65.

[55] *Summa de anima* II, 37, *ed. cit.,* pp. 290-94.

Taking his cue from Avicenna, Albert teaches that one must distinguish what the soul is in itself from what it is in relation to the body.[56] He insists that the soul is a complete spiritual substance, differing from an angel only in having a natural inclination to be united to the body. But as to whether it may therefore be considered a *hoc aliquid*, he says different things, even in the same work. In distinction 1 of book 2 of his commentary on the *Sentences*, he expands the definition of *hoc aliquid* to be not only "that which consists of matter and form, but that which consists of potency and act."[57] Spiritual bodies that lack matter can nevertheless be *hoc aliquid* because "in them is *quod est* and *quo est*, of which neither can ever be separated from the other, so that *quod est* might bespeak 'this something' that truly exists in nature, [and] *quo est* might bespeak the very principle of understanding and subsisting in such a being."[58]

But in distinction 17, he has second thoughts, since Aristotle had restricted the use of *hoc aliquid* to substances composed of matter and form, and Albert denied that the soul was so composed. "The soul is called a *hoc aliquid* by the *magistri*," he says, "but not by the philosophers or saints."[59] His objection is based on what Aristotle says at the beginning of book 2 of *De anima*, that neither matter nor form nor the soul is a *hoc aliquid*. The soul is a composite substance, to be sure, but it is not composed in the manner of a *hoc aliquid* (that is, of matter and form). "According to nature, it has a dependence on the body, although it is still able to exist without it." But a *hoc aliquid* is a "forma contracta per materiam."[60]

It is clear from this that Albert denies the matter-form composition of the soul (which the term *hoc aliquid* would imply), but he must still explain how the soul is an individually existing substance. Like John of La Rochelle and

---

[56] See Albert-M. Ethier, "La double définition de l'âme humaine chez St. Albert le Grand," *Études et Recherches publiées par le Collège dominicain d'Ottawa* I, Philosophie; Cahier I, pp. 79-110.

[57] "Unde dico non solum esse *hoc aliquid*, quod est ex materia et forma, sed quod est ex potentia et actu." *Comm. in Sent.* II, d. 1, art. 4, sol., *Opera* 25, p. 8.

[58] "Spiritualium autem quae sunt *hoc aliquid*, nulla est materia meo judicio, sed in ipsis est *quod est* et *quo est*: quorum neutrum nunquam separatur ab altero, ut *quod est* dicat hoc aliquid quod vere est in natura, *quo est* dicat principium intelligendi et subsistendi in tali esse." *Loc. cit.*

[59] "... quod anima sit hoc aliquid, hoc est dictum a Magistris, sed non a Philosophis nec a Sanctis." *Comm. in Sent.* II, d. 17 C, art. 2, ad 2. *Opera* 27, p. 299.

[60] *Summa de creaturis*, Pars I, tract. 1, q. 2, art. 2, ad 3. *Opera* 34, p. 325.

several others, he takes from Boethius the traditional distinction between *quod est* and *quo est*. *Quo est* he calls the form of the whole (*forma totius*), and *quod est* is the whole of which it is the form. "This composition," he says, "is in incorruptible and ungenerable things, in which the form of the whole does not differ from the form of matter, because it does not have matter." Gilson points out that it is very difficult, from what Albert says, to determine what the *fundamentum* of the soul is,[61] that is, to what does the form give form? And the *quod est* of his account comes very close to being the spiritual matter of Avicebron, whose existence Albert is at pains to deny. In fact, he is contemptuous of the *Fons vitae* and in his commentary on *De causis* doubts that Avicebron wrote it: "nec puto quod Avicebron hunc librum fecit, sed quod quidam sophistarum confinxerunt eum sub nomine suo."[62] If the soul were a simple form, it would not be able to exist apart from matter, and so would perish at the death of the body. If it were a complete substance composed of matter and form, it would not be able to enter into a further essential unity with the body.

Albert's way out is to avoid calling the soul the form of the body. He prefers to call it, as it is defined by Aristotle, the first act or perfection of the body. Much as he objected to the use of the term *hoc aliquid* for the soul, since this implies that it contains matter, he objects in the first place to the use of the term 'form' for the soul, since this implies that it can exist only in union with matter. A perfection or act, however, can exist apart from the thing it perfects, or of which it is the first act, just as a sailor can exist without a ship. The reason for his second objection is that "form bespeaks a comparison to that which is most remote from its complement, that is, the potency of matter. But perfection bespeaks a comparison to the perfection of a thing ... and since the soul is thus compared to the body, it is better called a perfection than a form."[63] This shows that Albert, like his contemporaries, continued to

---

[61] Étienne Gilson, "L'âme raisonnable chez Albert le Grand," *AHDLMA* 18 (1943), 5-72, on p. 40. On Albert's use of *quod est* as equivalent to matter, see A. Forest, *La structure métaphysique du concret selon saint Thomas d'Aquin* (Paris, 1931). pp. 124-25.

[62] *Comm. in De causis* I, tract. 3, c. 4. *Opera* 10, p. 407.

[63] "Ad aliud dicendum quod etiam duplex est ratio quare melius dicitur actus vel perfectio quam forma. Quarum una est, quia forma proprie secundum naturalem philosophiam est illa quae habet esse in hoc materia, et non est sine ea. Perfectio autem quaedam bene est sine perfecto secundum suam substantiam, sicut nauta sine navi. ... Secunda est, quia forma dicit comparationem ad id quod remotissimum est a complemento, hoc est, ad potestatem materiae. Perfectio autem dicit comparationem ad rem perfectam non tantum in materia, sed in omnibus quae exiguntur rei ... et cum sic anima comparetur ad corpus, melius

consider the soul and body as two complete substances and to consider the body, rather than matter, the subject of the soul.   As Anton Pegis aptly summed it up: "whereas *matter* would be the subject of form, *body* is the subject of soul as *actus* or *perfectio*."[64]

But even though the soul can exist in separation from the body, its nature is to be united to it. Body and soul have a mutual dependence on each other, and they will be united in the final resurrection. And it is because of this mutual dependence that they are able to constitute one substance.[65]

Albert resists the compound soul theory, as presented by Philip the Chancellor and maintained by others in the faculty of arts:

> For the rational soul has within itself the vegetative and sensitive faculties, according to which it is the act of the body. For just as a three-sided figure is within a four-sided figure potentially and not through its own proper being, thus the vegetative and sensitive are in the rational but do not have the being of the species of the rational soul. But this does not mean that if the rational soul is the act of the body, each of its faculties is joined to an organ. ... The rational soul, in the body of which it is the species, is not joined to organs according to all its parts. Moreover, the vegetative and sensitive are not in the rational as species, but as faculties.[66]

---

dicitur perfectio quam forma." *Summa de creaturis*, Pars II, tract. 1, q. 4, art. 1, ad 6. *Opera* 35, p. 35.

[64]   Anton Pegis, *St. Thomas and the Problem of the Soul*, p. 96.

[65]   "[C]orpus bene separatur ab anima ut materia, sed non ut actu manens ut subiectum separatur ab accidente. Similiter anima rationalis dependentiam habet ad corpus, eo quod est unibilis ei et unitur ei in resurrectione novissima. Et ideo cum sic utrumque dependeat ad alterum ex eis fit unum per substantiam." *Summa de creaturis*, Pars II, tract. 1, q. 4, art. 5, ad 5. *Opera* 35, p. 53.

[66]   "Anima enim rationalis habet in se potentiam vegetabilem et sensibilem, secundum quas est actus corporis: sicut enim trigonum est in tetragono potentia et non per esse proprium, sic vegetativum et sensitivum sunt in rationali potentia non habentia esse speciei rationalis animae. Non tamen oportet, quod si anima rationalis est actus corporis, quod quaelibet potentia ejus sit affixa organo. ... [R]ationalis in illo corpore cujus est species, non secundum omnes partes conjungitur organis. Vegetativum autem et sensitivum non sunt in rationali ut species, sed ut potentiae." *Summa de creaturis*, Pars II, tract. 1, q. 4, art. 1, ad 6. *Opera* 35, p. 35.

He goes on to attack Averroes's position on this matter, and this is the view of Albert's that Siger will single out to argue against:

> And many have been deceived on this matter, of whom the first is the Commentator himself, Averroes, who says that this definition through prior and posterior is suited to the soul. It is obvious that this is not true, unless one posits in man three perfections; which is impossible because perfected things differ according to perfections. Whence if there were three perfections in man (i.e., vegetative, sensitive, and rational), man would not be one.[67]

He also opposed Philip the Chancellor's position that the rational soul is united to the body mediately:

> But others have said that the vegetative and sensitive are immediately the perfections of the body to which they are united, but the rational is a mediated perfection because it is joined to the body in man only by the mediation of the vegetative and sensitive. But if it were thus, then it would have to be that by which man is defined and which gives him being in species and *ratio*, and it would not be his unmediated act; which is against the consensus of philosophy and against reason, because nothing is thus unmediated to a thing such as that by which it is its being and perfection. And in man this is neither the vegetative nor the sensitive, but the rational.[68]

---

[67] "Unde de hoc decepti sunt plurimi, quorum primus est ipse Commentator Averroes, qui dicit, quod haec diffinitio per prius et posterius aptatur animae. Quod patet non esse verum, nisi poneretur in homine tres perfectiones, quod est impossibile. Quia secundum perfectiones differunt perfecta. Unde si tres perfectiones essent in homine, vegetativa, sensibilis, et rationalis, non esset homo unus." *Ibid.*, ad 7, p. 35.

[68] "Alii vero dixerunt, quod vegetabile et sensibile sunt immediate perfectiones corporis cui uniuntur: sed rationalis est perfectio mediata, eo quod non nisi mediantibus vegetabili et sensibili conjungitur corpori in homine. Sed si hoc esset, tunc oporteret, quod illud a quo diffinitur homo, et quod dat ei esse in speciem et rationem, non esset immediatus actus: quod est contra communem philosophiam et rationem: quia nihil est ita immediatum rei sicut id a quo est esse et perfectio ipsius: hoc autem in homine non est vegetabile, neque sensibile, sed rationale." *Loc. cit.*, p. 36.

If the rational soul is indeed the perfection of the body and is joined to it immediately, but is nevertheless a complete substance and not strictly speaking
the form of the body, there is a problem in accounting for the existence of the
body of which it is the perfection and to which, he says, it gives being and
ratio, since he also teaches that the body, and not prime matter, is the subject
of the soul. Theodore Crowley has understood Albert as agreeing with Roland
of Cremona on the development of the embryo prior to the infusion of the
rational soul,[69] but Albert's meaning is more implicit than explicit. In the
Summa de creaturis, responding to the question of whether the soul is in the
semen, Albert reasserts his position that the soul is simple: "For we say that in
man there is one substance that is the soul, and not three substances. And
therefore we cannot say that a substance is partly created by God in its rational
powers and partly educed from the substance of the semen in its sensible and
vegetative powers; but the whole is created by God and infused into the
body."[70] And in his *Summa theologiae* he takes up the same problem: "The
same substance is not corruptible and incorruptible, for that substance that
thus prepares [the body] and is corruptible is not the same as the substance of
the rational soul. Moreover, the sense faculty that Augustine says is separable
with the body is of one substance with the intellective power and, although it
is not separable by reason of the fact that it is sensible, it is nevertheless
separable by reason of the substance in which it is founded."[71] Albert
clarifies his position in his later work, *De natura et origine animae*, in which
he adopts a position much like that of Aquinas on the progression of forms in
the developing embryo. Generation, he says, is a continuous process of that
which is in potency coming forth into act, until the process is terminated in its
ultimate fulfilment. Continuous time measures all motion, and the process of
generation is a succession of form after form, and in every becoming there are

---

[69]   *Roger Bacon*, p. 133.

[70]   "Nos autem dicimus quod in homine est una substantia que est anima, et non tres
substanti-ae, et ideo non possumus dicere, quod ex hoc substantia in parte creatur a Deo
quantum ad rationales potentias, et in parte educatur ax substantia seminis quantum ad
potentias sensibilies et vegetabiles; sed totum dicimus creari a Deo, et infundi corpori."
Summa de creaturis 2, q. 17, ad 22, *ed. cit.*, p. 159.

[71]   "[N]on est eadem substantia corruptibilis et incorruptibilis. Substantia enim illa,
quae ita praeparat et est corruptibilis, non est eadem cum ipsa substantia animae rationalis;
virtus autem sentiendi, quam dicit Augustinus separari cum corpore, unius substantiae est cum
potentia intellectiva et, licet non sit separabilis ex ratione qua est sensibilis, est tamen
separabilis ratione substantiae in qua fundatur." *Summa theologiae* 2, n. 332, ad 2, *ed. cit.*, p.
404.

infinitely many intermediate stages. The generation of substantial forms is not divided by alteration, because the first qualities informed by the substantial forms are instruments of alteration, by which matter is brought forth from potency into act.[72]   But the rational soul is not engendered by this process, for "it is not the act of any body, nor is it a corporeal form, nor a power acting in a body."[73] The rational soul is the one substance of man, but it collects in itself the powers of all the forms preceding it in the order of nature. These are fully contained in the intellectual nature, as in its final terminus. But there will only be one form at a time. "The whole substance of the soul is joined to the faculties of the vegetative, and the sensitive is joined to the bodily organs, and it does not act without them. But to the powers of the final complement, the rational and intellectual, is joined absolutely no part of the body, nor is it the act of any part of the body."[74]

Albert made great progress in getting at Aristotle's meaning of the agent and possible intellects. Concerning the agent intellect, after reciting the range of opinions, from those who deny its existence to a very full account of Avicenna's theory, Albert says:

> But I do not accept any of these opinions. Following
> Aristotle and Averroes ... I say that the human agent
> intellect is joined to the human soul, and is simple, and does
> not contain intelligibles, but that it itself acts on the possible
> intellect from the phantasms, as Averroes expressly says in
> his comment on *De anima.*[75]

---

[72]   *De natura et origine animae* 1, 4. *Opera* 9, pp. 386-87.

[73]   "... ipsa non est actus alicujus corporis, nec est forma corporalis, neque virtus operans in corpore." *Op. cit.* 1, 5. *Ed. cit.*, p. 389.

[74]   "... ipsa etiam tota substantia animae viribus potentiarum vegetabilis et sensi-bilis conjungitur organis corporalibus, et non agit sine illis, sed viribus complementi ultimi rationalis et intellectualis, nulli omnino parti corporis conjungitur, nec est actus alicujus partis corporis." *Op. cit.* 1, 6, *ed. cit.*, pp. 391-92.

[75]   "Sed nos nihil horum dicimus: sequentes enim Aristotelem et Averroem ... dicimus intellectus agentem humanum esse conjunctum animae humanae, et esse simplicem, et non habere intelligibilia, sed agere ipsa in intellectu possibili ex phantasmatibus, sicut expresse dicit Averroes in commento libri de anima." *Summa de creaturis*, Pars II, tract. 1, q. 55, art. 3, sol. *Opera* 35, p. 466.

It is called separated "because it is not the act of any part of the body. ... But it is nevertheless part of the intellective soul, because the intellective soul is the act of the body in such a way that it does not perform its operations in an organ according to any of its parts."[76] The soul knows some things by virtue of the influence of a higher light that descends on the soul from the separated Intelligences, but this is not an action of the agent intellect, which is not an instrument of the Intelligences. It acts through its substance and not through some species of intelligibles that it has within itself.

And from Averroes, Albert takes a position that will later be used as a stock argument by the 'heterodox Aristotelians' against the argument that if there were one intellect for all men, all men would simultaneously know the same thing: "The diversity of action of the agent intellect is not from the agent intellect, but from the phantasms. And this is what Averroes says in his comment on book 3 of *De anima*,"[77] which Albert then quotes. But of course Albert was confining his remarks to the agent, not the possible, intellect, and he was assuming that both Aristotle and Averroes considered it to be a part of the individual human soul. The agent intellect, according to its substance, says Albert, "is the potency and active principle of intelligibles; and for this reason the Philosopher says that it is the intellect *quo est omnia facere*. By itself, separated from the possible intellect, it understands nothing."[78]

The agent and possible intellects in the soul come about from the diversity of principles of which the soul is composed. These "principles are *quod est* and *quo est*, or act and potency, if we understand these words broadly. And for this reason I say that the agent intellect is the part of the soul flowing from *quo est*, or act; and the possible is the part of the soul flowing from *quod est*, or potency."[79]

---

[76] "[I]ntellectus agens dicitur *separatus*, eo quod non est actus alicujus partis corporis. ... Nihilominus tamen est pars animae intellectualis; quia anima intellectiva sic est actus corporis, quod non secundum quamlibet partem ejus se habet in organo." *Ibid.*, ad 1, p. 466.

[77] "...diversitas actionis intellectus agentis non est ex intellectu agente, sed ex phantasmate: et hoc est quod Averroes in commento super tertium de *Anima* dicit." *Ibid.*, ad 14, p. 467.

[78] *Summa de creaturis*, Pars II, tract. 1, q. 55, art. 5, sol. *Opera* 35, p. 473.

[79] "... quae principia sunt quod est et quo est, vel actus et potentia, si elargato nomine sumantur. Et propter hoc dicimus, quod intellectus agens est pars animae fluens ab eo quo est, sive actu: possibilis autem pars animae est fluens ab eo quod est, sive potentia." *Ibid.*, art. 3, sol., p. 470.

The possible intellect, Albert says,

> is a passive potency, if 'passion' is construed broadly to
> apply to both the physical and non-physical, as it is defined
> in book 5 of the *Metaphysics*, that is, that passive potency is
> the principle of transmutation from something other
> according to which it is other; for the possible intellect is
> changed by the action of another, which is the agent
> intellect, and its transmutation is the going forth from the
> potency of understanding to the act of understanding. And
> in this sense of passive potency it is defined by the
> Philosopher, through the definition given above.[80]

Although the intellective soul is free of matter, nevertheless the possible
intellect is founded on *quod est*, and it has a potency similar to matter.
Although the possible intellect is not composite, "it is itself the potency of the
composite intellective soul, and because of it the intellective soul is made the
subject of intelligibles, just as a thing composed of form and matter underlies
accidents not by virtue of form, but by virtue of matter."[81] With respect to its
substance, the possible intellect is incorruptible, although it is corruptible *per
accidens*, that is, through the act that is in it. But this corruption is nothing but
the forgetting of intelligibles.[82]

Albert was too good an Aristotelian to hold that the soul was at once a
complete substance and the form of the body. He solved the problem by
refusing to call it a form, preferring instead to call it the act of perfection of the
body. He continued to maintain that it was a complete substance, lacking
matter but having something analogous to matter, namely *quod est*. Although
he clearly teaches the unity of the soul, he continues to hold to a doctrine of
duality of substances: the soul is one, the body another; but each has a natural

---

[80] "[P]ossibilis intellectus est potentia passiva, si passio large sumatur ad physicam
et non physicam, secundum quod diffinitur in quinto *Metaphysicae*, scilicet quod potentia
passiva est principium transmutationis ex alio secundum quod est aliud: possibilis enim
intellectus transmutatur ex alio quod est intellectus agens, et sua transmutatio est exitus de
potentia intelligendi ad actum intelligendi: et in hac ratione potentiae passivae diffinitur a
Philosopho per superius datum diffinitionem." *Ibid.*, q. 56, art. 1, sol.

[81] "... licet possibilis intellectus non sit compositus, tamen ipse est potentia animae
intellectivae compositae: et gratia sui anima intellectiva efficitur subjectum intelligibilium,
sicut et compositum ex forma et materia substat accidentibus, non gratia formae, sed gratia
materiae." *Ibid.*, ad 2, pp. 478-79.

[82] *Ibid.*, art. 4, sol., p. 483.

dependence on the other and is unible to the other. But in spite of Albert's attempt to explain how they constitute one substance, the nature of their union is essentially operational or instrumental, not substantial. He successfully avoided making the soul a material form, even though it operated partly through matter, and so he was able to save its immortality and its ability to function without using a bodily organ, as in understanding. But human unity is still quite tenuous in his thought. In one passage, Albert actually admits that if the soul were only a substance and not also an *actus* in relation to the body, it would not be able to be joined to the body in such a way as to constitute a genuine unity with it.[83]   And this bifurcation of his thought is clearly stated by him in his summing up of his thought on the soul in the *Summa theologiae*: "When we consider the soul according to itself, we shall agree with Plato; but when we consider it according to the form of animation that it gives to the body, we shall agree with Aristotle."[84]

---

[83] "Ad aliud dicendum, quod haec propositio est falsa si universaliter ponatur: Omnis motor secundum locum processivum non solum est actus, sed actu ens per se ipsum: sensibilis enim anima est motor processive secundum locum in brutis, et tamen non est actu ens per seipsam, sed est actus corporis solum. Si vero non universaliter, sed de anima rationali tantum proponitur, tunc supra solutio est argumentatio, ubi distinctum est, quod anima rationalis non tantum est substantia per se ens, nec tantum actus, sed substantia et actus: sed si esset substantia per se tantum, non esset unibilis alteri per constitutionem unius per substantiam." *Summa theologiae*, Pars I, tract. 12, q. 69, *ed. cit.* 33, pp. 49-50. See Pegis, *St. Thomas and the Problem of the Soul*, pp. 101-05.

[84] "... animam considerando secundum se, consentiemus Platoni; considerando autem secundum formam animationis quam dat corpori, consentiemus Aristoteli." *Summa theologiae*, Pars I, tract. 12, q. 69, ad 2, *ed. cit.* 33, pp. 13-14.

CHAPTER FIVE

A POLARIZATION OF VIEWS

During the decade of the 1250s, thanks largely to the thought of two men, Bonaventure and Thomas Aquinas, a kind of order, or more accurately a polarization of views, was imposed upon the study of the rational soul. While both men were highly original in their solutions, both were deeply indebted to the preceding half-century of thought on the subject. But each made a vastly different selection from that tradition, Bonaventure developing the Semitic sources and Aquinas the Greek, especially Aristotle. Although both men had a common goal—to safeguard the immortality of the soul while still maintaining the unity of the human being—they went about it in quite different ways. And although they used a common technical philosophical vocabulary, they meant radically different things by the terms they used.

BONAVENTURE. Like all writers of his generation, Bonaventure was much influenced by Avicenna, especially in holding that the soul should be considered from two points of view: in itself, according to its substance, it is the body's perfection; and with respect to its powers, it is the body's mover. Like Albert the Great, Bonaventure was uncomfortable with the term 'form' for the soul, preferring instead to call it the body's perfection or mover,[1] but he did, in the final analysis, say that it was the form of man, united to the body substantially and not accidentally.

But unlike Albert, he insisted that the soul was a *hoc aliquid*. Still, in Bonaventure's thought we find the extreme dualism that was so pronounced a feature of the teaching of Alexander of Hales and William of Auvergne to be much attenuated. For although the soul is a *hoc aliquid*, its nature is to be a part of man and not the whole man.[2] "The completion of nature," he says,

---

[1] "Quoniam autem ut beatificabilis est immutabilis; ideo, cum unitur mortali corpori, potest ab eo separari; ac per hoc non tantum *forma* est verum etiam *hoc aliquid*; et ideo non tantum unitur corpori ut *perfectio,* verum etiam ut *motor;* et sic *perficit* per essentiam, quod *movet* pariter per potentiam." *Breviloquium* 2, 9, 5, *Tria Opuscula Seraphici Doctoris S. Bonaventurae: Breviloquium, Itinerarium Mentis in Deum, et De reductione Artium ad Theologiam* (4th ed., Quaracchi, 1925), p. 84; and "[C]orpus unitur animae ut *perficienti* et *moventi* et *ad beatitudinem sursum tendenti.*" *Ibid.* 2, 10, 4, *ed. cit.,* p. 88.

[2] "...totus homo componitur ex carne et anima." *Comm. in Sent.* 3, d. 21, art 1, q. 3, fund. 3, *Doctoris Seraphici S. Bonaventuae Opera Omnia* (10 vols., Quaracchi, 1882-1902),

"requires that man consist at once of body and soul, just as of matter and form, which have a mutual appetite for and inclination toward each other."[3] Neither the soul nor the body enjoys its complete existence apart from the other, and although the soul (the formal principle) has the capacity for existing after the dissolution of the body (the material principle), its full actuality consists of being united to the body.[4] This unibility is an essential and permanent aspect of its nature.[5] The separated soul is always a true substance but never a complete substance.[6]

In spite of his granting the soul's incomplete state when separated from the body, Bonaventure found it necessary to insist that is was a *hoc aliquid* and therefore composed of form and matter—not corporeal but spiritual matter. In developing his view of spiritual matter, although he was obviously indebted, directly or indirectly, to Avicebron, he cited Augustine's *Confessions* 12, 3, 6 as his authority. Matter considered in itself, he says, is the possibility of taking on every kind of form. But in nature it never exists in this condition and is, in fact, never separated from some form or other. Considered in itself it is neither corporeal nor spiritual, and its essence is indifferent to either. It acquires either spiritual or corporeal being by virtue of subsequent forms, which make it one or the other of these. In the soul, it is (spiritual) matter that both stamps it a creature that has received existence (from God) and makes possible its subsistence. In confirmation of this he cites Boethius to the effect that no creature is pure act, because *quod est* and *quo est* differ in it, so that in every creature there is both act and possibility.[7]

3, p. 440. "[D]e integritate humanae naturae est non tantum anima, verum etiam caro." *Comm. in Sent.* 3, d. 21, art. 1, q. 2, *fund. 1, ed. cit.* 3, p. 438.

[3] "*Completio* vero *naturae* requirit ut homo constet simul ex corpore et anima tanquam ex materia et forma, quae mutuum habent appetitum et inclinationem mutuam." *Breviloquium* 7, 5, 2, *ed. cit.* 5, p. 269.

[4] See the discussion in Edouard Szdzuj, O.F.M., "Saint Bonaventure et la problème du rapport entre l'âme et le corps," *La France Franciscaine* 15 (1932), 283-310, on p. 293.

[5] "Ille enim animae appetitus est ratione perfectionis *esse naturalis*, sed iste solum est quantum ad *be-ne esse*; et ideo ille non potest terminari, nisi corpus uniatur." *Comm. in Sent.* 4, d. 49, pars 2, sect. 1, art. 3, q. 1, ad 2, *ed. cit.* 4, p. 1019. See also *Breviloquium* 7, 7, 4, *ed. cit.* 5, p. 279.

[6] E. Szdzuj, *art. cit.*, p. 294.

[7] "Nulla creatura est actus purus, quia in omni creatura, ut dicit Boethius, differt *quo est* et *quod est*; ergo in omni creatura est actus cum possibili; sed omnis talis habet in se

The rational soul, since it is a *hoc aliquid*, and naturally subsists by itself, and acts and undergoes, moves and is moved, has within itself the foundation of its own existence: both a material principle by which it exists, and a formal principle by which it has its being. One need not say the same about the soul of beasts, since it is founded in the body. Therefore, since the principle by which the existence of a creature is fixed in itself is a material principle, it must be conceded that the human soul contains matter. But this matter is above the condition of extension and above the condition of privation and corruption, and therefore it is called spiritual matter.[8]

Bonaventure also took from Augustine the Stoic concept of the *rationes seminales*, which bestow on matter a kind of determinative potency. So matter is not simply the ground of subsistence, but it possesses within itself that out of which form can be produced by an agent.[9] Thus the distinction between matter and form in Bonaventure's thought is blurred.

Since Bonaventure did call the soul the form of man, he was faced with explaining how two substances, body and soul, could constitute a genuine unity. In distinction 17 of book 2 of his *Sentences* commentary, he presents among his preliminary arguments a very strong one for considering the soul to be pure form:

Everything that has matter and form as constitutive parts is a *hoc aliquid* and a complete being. But nothing that is a *hoc aliquid* and a complete being can enter into the constitution of a third being. But the rational soul enters

---

multiformitatem et caret simplicitate." *Comm. in Sent.* 1, d. 8, pars 2, art. unicus, q. 2, *ed. cit.* 1, p. 167.

[8] "[A]nima rationalis, cum sit *hoc aliquid* et per se nata subsistere et agere et pati, movere et moveri, quod habet intra se *fundamentum* suae existentiae et *principium materiale,* a quo habet *existere,* et *formale,* a quo habet *esse.* De brutali autem non oportet illud dicere, cum ipsa fundetur in corpore. Cum igitur principium, a quo est *fixa existentia* creaturae in se, sit principium materiale; concedendum est, animam humanam materiam habere. Illa autem materia sublevata est supra *esse* extensionis, et supra *esse* privationis et corruptionis, et ideo dicitur materia spiritualis." *Comm. in Sent.* 2, d. 17, art. 1, q. 2, conclusio, *ed. cit.,* 2, pp. 414-15.

[9] See the extensive discussions of this in A. C. Pegis, *St. Thomas and the Problem of the Soul,* pp. 44-50 and Étienne Gilson, *La Philosophie de S. Bonaventure* (Paris, 1925), ch. 10.

into the constitution of a third being, so that from soul and
body is made one thing through essence. Therefore the
rational soul is not a *hoc aliquid*. Therefore it is either
matter or pure form. It is not matter; therefore it is form.[10]

This is indeed the crux of the problem. Bonaventure solved it by applying an
essentially Avicennan notion of matter and form to the body-soul problem,[11]
basing his argument upon the incomplete nature of either body or soul without
the other. The above argument, he says, only holds where the matter and
form joined to each other mutually exhaust the other's appetite for fulfilment.
In such cases, matter and form constitute a complete being. But it is not thus
with the soul and the body. Each is composed of its own matter and form, but
in neither case is the appetite of either one exhausted by its own matter and
form; each has a remaining appetite to be joined to the other, the soul to
perfect the body, the body to be perfected by the soul. It is in the composite
that each one finds its highest development. The soul is not imprisoned in the
body but has a powerful natural inclination to be united to it. It is in this sense
that the soul is the substantial form of man and not just an accidental form.[12]
But the body, as body, has its own form. In treating the question of the unity
of the soul and the relationship of the rational soul to the sensitive and
vegetative souls of the embryo, Bonaventure's final position seems to have
been much the same as that reported by Richard Rufus, and apparently
maintained by Alexander of Hales and explicitly by John of La Rochelle, and
indeed was very possibly Rufus's source: that is, the double vegetative and
sensitive soul in man. Since Bonaventure was unwilling to depart from what
he considered to be the teaching of Augustine, and since he considered *De
spiritu et anima* and *De ecclesiasticis dogmatibus* to be authentic Augustinian
works, Bonaventure held to the view that after the rational soul is infused, it
performs all the vital functions of man. But instead of following Philip the
Chancellor, Roland of Cremona, or Peter of Spain, he settled on quite a

---

[10] "Item, omne quod habet materiam et formam ut partes constitutivas, est *hoc aliquid*
et est *completum*; nihil autem, quod est *hoc aliquid* et *completum* in se, venit ad
constitutionem tertii; sed anima rationalis venit ad constitutionem tertii, ita quod ex anima et
corpore fit unum per essentiam; ergo anima non est *hoc aliquid;* ergo vel *materia*, vel *forma
pura*; non *materia*, ergo *forma*." *Comm. in Sent.* 2, d. 17, art. 1, q. 2, obi. 6, *ed. cit.*, 2, p. 413.

[11] See Szdzuj, *art. cit.*, p. 297.

[12] "Hoc enim, quod est animam uniri corpori humano sive vivificare corpus humanum, non dicit actum *accidentalem* nec dicit actum *ignobilem*: non *accidentalem,* quia ratione
illius est anima forma substantialis; non *ignobilem*, quia ratione illius est anima nobilissima
formarum omnium, et in anima stat appetitus totius naturae." *Comm. in Sent.* 2, d. 1, pars 2,
art. 3, q. 2, concl., *ed. cit.*, 2, p. 50.

different solution. He attempts to uphold both the unity of the soul (which, as we have seen, he calls the substantial form of man) and an implicit theory of plurality of forms in each existing substance, the last form supplying the final perfection of the substance. But he denies that as a perfecting principle the soul is joined to the body by any medium. The soul by itself perfects the human body, just as form is united to matter by itself.[13]

This doctrine of the double vegetative and sensitive soul does not appear in the Quaracchi edition of Bonaventure's *Sentences* commentary , but F. -M. Henquinet in 1932 published a description of an autograph draft of Bonaventure's for portions of the *Sentences* commentary,[14] in which he explicitly maintains it:

> The fifth position is that there is a double vegetative and sensitive soul in man: one that operates before the complete being of man, and this disposes the body for receiving the rational soul and ceases its operation when the rational soul arrives and remains as a disposition and medium of union; and the rational soul, which has three powers, namely vegetation, sense, and reason, is infused. And by thus holding to this path, we are in accord with the philosophers and the saints; and thus one can reply to both sides.[15]

---

[13] "... uniri *corpori* non est proprium animae rationalis; sed tamen unire *corpori humano,* sicut dicit illud quod est animae essentiale et nobile, sic etiam importat quod est proprium. — Et per hoc patet aliud. Nam illud quo mediante anima perficit corpus humanum, est illud quo anima est anima rationalis, et quod etiam est principium aliarum nobilium operationum; sed perficit se ipsa; se ipsa enim anima perficit corpus, sicut forma se ipsa unitur materiae." *Comm. in Sent.* 2, d. 1, pars 2, art. 3, q. 2,concl. 2.3, *ed. cit.,* 2, pp. 50-51.

[14] F. -M. Henquinet, "Un brouillon autographe de S. Bonaventure sur le commentaire des Sentences," *Études Franciscaines* 44 (1932), 633-55.

[15] "Quinta positio est quod duplex est vegetabilis et sensibilis in homine, una quae operatur ante completum esse hominis et haec disponit corpus ad susceptionem rationalis, et cessat eius operatio adveniente rationali et manet tamquam dispositio et medium uniendi et infunditur anima rationalis quae habet illas tres potentias, scilicet potentia vegetandi, sentiendi, et ratiocinandi. Et sic hanc viam tenendo concordamus philosophis et sanctis et sic potest ad utramque partem responderi." This paragraph was transcribed from Assisi MS Bibl. com. 186, fol. 97d by F. -M. Henquinet at the request of Theodore Crowley, who printed it in *Roger Bacon,* p. 130, n. 39. The pertinent section of the *Collationes in Hexameron* as edited by Ferdinand Delorme, O.F.M., *S. Bonaventurae Collationes in Hexaemeron et Bonaventuriana Quaedam Selecta,* (Florence/Quaracchi, 1934) is printed by Crowley, pp. 130-32 as follows: "Si loquamur de potentia quae est actus incompletus, ut est ratio seminalis, quae secundum naturalem philosophum dicitur potentia activa et est aliquid formae, hic actus super potentiam alicubi addit modum essendi, alicubi partem essentiae, non alterius, sed ordinabilis cum illo. Quod patet, quia animatum addit vere super corpus, sensibile super animatum,

Hence the vegetative-sensitive soul of the embryo ceases to function as a soul but continues to exist after the infusion of the rational soul as a "disposition and medium of union." It is not clear how he would reconcile this with his earlier contention that the soul is united to the body without any medium.

Theodore Crowley has pointed out that this position is consistent with what Bonaventure says in the *Collationes in hexameron* (especially the Delorme edition) but not with what appears in the *Opera omnia* edition of the *Sentences* commentary; and he aptly concludes: "It appears, therefore, that the Assisi codex contains questions that are posterior to the commentary of the Quaracchi edition—unless, of course, we are prepared to admit that St. Bonaventure changed his mind twice on the question of the soul."[16]

Bonaventure's teaching on the intellect owes something to his Parisian predecessors, much to Robert Grosseteste, and most to the integrative and critical powers of his own mind. When his views are considered apart from the larger context of his thought, they appear to be very similar to those we have encountered among masters of diverse stamps during the preceding twenty years, including Alexander of Hales, John of La Rochelle, and especially Peter of Spain; and although Bonaventure is less Avicennist than Peter, he is surprisingly more Aristotelian. The overall model of his doctrine is based on the two faces of the soul, and he considers the soul under a threefold aspect: toward things below it, toward itself, and toward that which is above it, that is, God. He considers the agent intellect to be a power of the individual soul but holds that the knowledge thereby obtained is insufficient. And he teaches a form of attenuated innatism.

---

rationale super sensibile. Si ergo loquamur de hac additione, hic est additio realis, quia anima est aliquid ultra naturam corporis, licet essentialiter ad illud ordinatum; similiter de sensibilitate et rationabilitate, ut tetragonum super trigonum addit novum angulum et pentagonum super tetragonum addit non solum modum essendi, ut patet II *De anima*, in principio. Si autem loquamur de potentia quae solum est passiva, ut quod modo est in potentia, modo in actu, addit modum essendi solum. Hic nota quod hoc non est verum quod ultima forma addatur primae materiae nulla forma interiecta, quia sensibile et vegetabile interponuntur et illa etiam additio est realis et naturalis; nec tamen de vivo potentia fit actu vivum: radicale vivum erat ibi" (*Visio* I, coll. I, 10, pp. 53-54). "*Observa iustitiam*; sicut appetitus materiae facit eam habilem ad talem vel talem formam, non quod dispositiones illae sint immediatae, nec perimuntur per dispositiones medias; nam adveniente ultima forma non destruuntur dispositiones mediae sive antecedentes, sed complentur, ita quod sit una forma unius perfectibilis. Unde nec in uno homine debent dici tres animae, vegetabilis, sensibilis, rationalis, sed una anima cum suis potentiis, differentibus quidem per essentiam, licet cum anima conveniant secundum substantiam et subiectum. Et si inveniatur ab Aristotele dictum quod sint tres animae, intelligitur in diversis" (Principium, Coll. II, I, 2, p. 20).

[16]   *Roger Bacon*, p. 133.

Like Grosseteste he denies to philosophy any sphere of independence; and although I have not found the formula *aspectus et affectus mentis* among his writings, he has integrated its implications into the very core of his thought. Also, he bases his theory of human cognition upon the fact that the soul is the highest imitative likeness of the Trinity. And he stresses the mutual dependence of abstractive knowledge gained from sense experience and divine illumination.

At its lowest level, human cognition in Bonaventure's thought is quite a pure Aristotelianism. The possible intellect receives the intelligible species provided by the phantasms, and they are then illuminated by the agent intellect. The agent intellect is the *vis activa* given by God to the soul that enables it to discern the intelligibility of sense data.[17] Bonaventure denies that the possible and agent intellects are distinct powers; they are rather two operations of one power of the human soul.[18]

This type of abstractive knowledge from sense data is essential for human knowledge;[19] without it we would know nothing, for it determines and specifies what we know. But it is insufficient, for at its best it provides only *scientia* and not *sapientia*, since it is of mutable creatures by mutable creatures. The soul, in its higher powers, which are most truly God's image in us, has three operations, all of which are independent of sense experience. Toward what is below it (the body) it is the ruler and administrator and the seat of judgment. Also, it knows itself, its states and its powers. And it turns itself to God and the eternal exemplars of things.[20] Because of man's status as the image of God, God cooperates with him in knowing, shedding his

---

[17] See the discussion in Otto Keicher, "Zur Lehre des ältesten Franziskanertheologen vom 'Intellectus Agens'," *Abhandlung aus dem Gebiete der Philosophie und ihrer Geschichte* (Freiburg i. B., 1913), pp 181-82 and Leonard J. Bowman, "The Development of the Doctrine of the Agent Intellect in the Franciscan School of the Thirteenth Century," *The Modern Schoolman* 50 (1973), 251-79, on p. 263.

[18] Bonaventure, *Comm. in Sent.* 2, d. 24, pars 1, art. 2, q. 4, concl., *ed. cit. p.* 570.

[19] "Unde licet anima secundum Augustinum connexa sit legibus aeternis, ... indubitanter tamen verum est, secundum quod dicit Philosophus, cognitionem generari in nobis via sensus, memoriae et experientiae, ex quibus colligitur *universale* in nobis, quod est principium artis et scientia." *Sermo* 4, 18, ed. Quaracchi 5, p. 572. See also Comm. in Sent 1, d. 16, art. unicus, q. 2, fund. 1: "Quod sit utilis ad innotescendum, videtur, quia cognitio nostra incipit a sensu; ergo si debemus elevari ad perceptionem intelligibilium, congruum et perutile est, quod aliquo modo praevia sit excitatio in sensu per signum." *Ed.cit.* 1, p. 281.

[20] See P. Bonifaz Anton Luyckx O.P., *Die Erkenntnislehre Bonaventuras.* BGPM XXIII, 3-4 (Münster i. W., 1923), pp. 171-85.

uncreated light, i.e., the eternal principle, as a regulative and motive law, although God is not to be considered the agent intellect properly speaking.[21] So sense knowledge and God's illumination are co-principles in the acquisition of certain knowledge. Neither is sufficient without the other. Abstraction specifies and determines the thing known; illumination certifies it. The nature of divine illumination in Bonaventure's thought is slightly ambiguous.[22] M. Hurley describes it as "something less than grace and something more than His ordinary cooperation."[23] Bonaventure says of it that "the eternal principle is the regulative and motive principle in all our certitude; it is necessarily present, not however by itself but with the created idea; it is contuited by us, not however in its full splendor, but only partly and as befits our present condition."[24]

Bonaventure's thought would provide the point of departure for a sizeable group of masters during the next generation. Among the characteristic doctrines of this group were: 1- plurality of forms; 2- universal hylomorphism; 3- the soul's being a *hoc aliquid*; 4- the two faces of the soul; 5- the simplicity of the soul (although they had some difficulty in reconciling this with a theory of plurality of forms). In addition, they added an insistence on the formula that the rational soul was the substantial form of the body. Bonaventure had used this phrase, but he had been conscientious in explaining precisely what he meant by it, and like Albert the Great he preferred the designation 'perfection.' But the debates against Averroes led many to stress the rational soul's being man's substantial form.

Despite the power of Bonaventure's thought, he seems to have had a blind spot regarding the soul, much as he did concerning the eternity of the world. On the latter question he denied the intelligibility of the concept of an

---

[21] "Donum scientiae duo antecedunt: unum est sicut lumen *innatum,* et aliud est sicut lumen *infusum.* Lumen innatum est lumen *naturalis iudicatorii* sive rationis; lumen superinfusum est *lumen fidei.* Quantum ad primum dicit: *Deus, qui dixit lucem splendescere,* id est lumen naturalis iudicatorii imposuit creaturae rationali, id est non solum intellectum possibilem, sed etiam intellectus agentem." Bonaventure, *De donis Spiritus Sancti, coll.* 4, 2, ed. Quaracchi, 5, p. 474.

[22] Its ambiguities are discussed by Bowman, *art. cit.*, p. 261.

[23] M. Hurley, "Illumination according to S. Bonaventure," *Gregorianum* 32 (1951), on p. 399.

[24] "Et ideo est *tertius modus* intelligendi, quasi medium tenens inter utramque viam, scilicet quod ad certitudinalem cognitionem necessario requiritur ratio aeterna ut *regulans* et *ratio motiva,* non quidem ut *sola* et in sua omnimoda claritate, sed cum ratione creata, et ut *ex parte* a nobis contuita secundum statum viae." *Quaestiones disputatae de scientia Christi,* Q. 4, concl., ed. Quaracchi 5, p. 23 . I have borrowed the translation from Hurley, *art. cit.*, p. 399

eternally created being, holding that 'eternal' and 'created' were contradictory. And concerning the soul, he could not see how it could be both individual and capable of life apart from the body unless it were a complete substance separate from the body and was composed of matter and form. His followers would exaggerate certain aspects of his thought, suppress others, and introduce an unmistakable element of personal animosity into the debate, which was quite inconsistent with Bonaventure's own character.

THOMAS AQUINAS. The central problem faced by the scholastics was how to reconcile the notion of the soul as a form with that of the soul as a complete spiritual substance. Some authors—Alexander Nequam, Alexander of Hales, William of Auvergne, and Albert the Great—denied in one way or another that it was a form; and those who did concede the use of this term for it always had to re-define 'form' in order to save the soul's substantiality and immortality. If it was the form of the body in any sense, the relation of the rational soul, which placed man in his species, to the vegetative and sensitive, became a major problem. If the rational soul itself was the only substantial form of a living human being and was the source of all the vital functions, then one had to account for the development of the embryo (which certainly had some kind of life) before the infusion of the rational soul. We have seen an interesting array of suggestions to account for this: William of Auvergne taught that the powers of the rational soul absorbed the weaker powers that had formed the embryo, and he called the sensitive soul the form of the body insofar as it was a body. Roland of Cremona and Peter of Spain had denied any pre-rational soul to the embryo, the former attributing its growth and development to the soul of the mother, the latter to powers inherent in the semen of the father. Several authors had suggested a progression of forms in which the succeeding one possessed everything that the prior one had plus something else; but they continued to consider the body an independent substance, the subject of the form 'soul.' More common were those who followed Philip the Chancellor in concocting some type of compound soul, including both corruptible and incorruptible constituents. The most important shortcoming of these explanations, even that of so subtle and able a thinker as Anonymous Vaticanus, was that the whole soul was not immortal; only its rational constituent was. Only Robert Grosseteste, it seems (and as we shall see below Anonymous Van Steenberghen), taught that the soul was infused at conception and was responsible both for the development of the embryo and for all of human life. Particularly troublesome was the doctrine of the form-matter composition of the soul: this guaranteed the soul's substantiality but made it virtually impossible to consider it the form of the body, since, as several masters pointed out, if it were already a complete substance it would be unable to enter into a further essential relationship with matter. Also, to an

Aristotelian, the rational soul had to be free of matter in order to perform its highest function, understanding. But it also had to have some sort of contact with matter in order to function as a soul. It would have to be a form of matter without being a material form, for in the latter case it could only exist in conjunction with the matter of which it was the form. It was at this stage of con-fusion that Thomas Aquinas entered the academic scene and tried to provide a solution at once theologically and philosophically sound.

In no aspect of Thomas Aquinas's thought is the originality and power of his mind more apparent than in his doctrine of the soul.[25] He saw the essence of the problem clearly: how to account for the genuine unity of man and at the same time save the subsistent reality of the soul and the non-material character of intellection. He treated this topic in a number of works ranging from his commentary on the *Sentences* of 1255 to the last period of his life, the *De unitate intellectus* of 1269. Much space in the *Summa contra gentiles* is devoted to it, he composed a separate commentary on Aristotle's *De anima*, and he touched on the subject in a number of other works. His teaching is consistent throughout his life, with only a shift of emphasis determined by the immediate nature and purpose of his various works.

In order to save the unity of man, Aquinas sacrificed the completeness of the soul. He did not consider the soul and body as two existing substances; rather they have a common principle of existence given to the substance 'man' by the soul, as form.[26] Although spiritual substances have a composition of form and existence, they must necessarily lack matter. Each one is a complete species and a *hoc aliquid*, subsistent and complete in its own nature. The soul, however, is a *hoc aliquid* only in being subsistent; it is not a complete species but only a part of man.[27] If it were indeed a complete substance, it would be unable to enter into a further composition in an essential way. But its nature is to be the form of the body, and it is only complete when it is so acting. It is ennobled and fulfilled by its relationship to the body of which it is the perfection. It is the composite of body and soul, not the soul alone, that

[25]   An old but still very useful treatment of Aquinas's doctrine of the soul is Anton Charles Pegis, *St. Thomas and the Problem of the Soul in the Thirteenth Century.*

[26]   See *Summa contra Gentiles* II, 56 and II, 69.

[27]   "Anima autem rationalis, quantum ad aliquid potest dici 'hoc aliquid,' secundum hoc quod potest esse per se subsistens. Sed quia non habet speciem completam, sed magis ei ut pars speciei, non omnino convenit ei quod sit hoc aliquid." *In De anima comm.* 215 (*De anima* 2, 11, 412a).

constitutes a person, and it is this personality in which human unity lies.[28]
This is, in a way, the same point that had been made by Grosseteste, but the
minds and procedures of the two men are so different that one can hardly
compare their views.

Even though the soul is not strictly speaking a *hoc aliquid*, it is a subsistent
being, capable of existing, even as an incomplete being, apart from the body.
The soul, like all spiritual substances, has a composition of *forma* and *esse*, or
quod est and *quo est*.[29] The potentiality in it refers only to existence, not to
both form and matter as in corporeal beings. It can therefore exist apart from
matter, since it does not depend on it, whereas in corporeal beings, form must
be joined to matter in order to exist.

But this account creates a serious problem concerning the soul's
individuation. If it is a spiritual substance, it therefore lacks matter, the
principle of individuation in Aquinas's thought. Other spiritual beings are
complete species. But the human soul is individual for each member of the
human race. Thomas's solution is that the soul is individuated by virtue of the
body (*ex corpore*), but not in the same way that material forms are
individuated by matter. Each soul is created for a particular body, and so,
even when it is separated from that body of which it is the form, it remains an
individual because, as a subsistent being, it will naturally retain its individual
perfection, just as it will retain its existence. And if it can exist by itself, there
is no reason why it should not continue to be what it has become through the
body.[30]

---

[28] "...quamvis anima sit dignior corpore, tamen unitur ei ut pars totius hominis, quod
quadammodo est dignior anima, inquantum est completior." *Comm. in Sent.* 3, d. 5, q. 3, art.
2 solutio et Ad 5.

[29] This is basic to Thomas's metaphysics and is often expanded in his works. See
especially *Comm. in Sent.* 2, d. 3, q. 1, art. 1, solutio and *De ente et essentia* 4. See also the
extended treatment in John Goheen, *The Problem of Matter and Form in the* De Ente et
Essentia *of Thomas Aquinas* (Cambridge, Mass., 1940).

[30] "Ad sextum dicendum quod, secundum predicta, in anima non est aliquid quo
individuatur, et hoc bene intellexerunt qui negaverunt eam esse hoc aliquid, et non quod non
habeat per se absolutum esse. ...Et dico quod non individuatur nisi ex corpore. Unde
impossibilis est error ponentium animas primo creatas et postea incorporeatas; quia non
efficiuntur plures nisi secundum quod iunguntur pluribus corporibus." *Comm. in Sent.* 1, d. 8,
q. 5, art. 2, ad 6. Also *De ente et essentia* 5: "Et licet individuatio eius ex corpore
occasionaliter dependeat quantum ad sui inchoationum, quia non acquiritur sibi esse
indiviuatum nisi in corpore cuius est actus, non tamen oportet quod, subtracto corpore,
individuatio pereat, quia cum habeat esse absolutum ex quo acquisitum est sibi esse
individuatum ex hoc quod facta est forma huius corporis, illud esse semper remanet
individuatum."

But by far the most original (and disconcerting to his colleagues) of Aquinas's teachings on the soul was his insistence on the unity of substantial form. It is form that gives being to a substance. Therefore one substance can have only one form, the form that confers being upon it. The subject of the soul is not the body, but prime matter.[31] The soul is the form of the body in the sense that it is the principle through which the body exists and is living.

But this brings us face to face with the problem that had bedeviled all of the scholastic writers on the soul: what was the relationship between the rational soul, created and infused into the developed body, and the principle which brought about the growth and organization of the embryo? For Thomas did not, like Grosseteste, hold that the rational soul was infused at conception and used only its lower powers successively until the embryo reached its full development as a human being. Instead he opted for a doctrine of succession of forms, similar in some ways but differing significantly in others, from other similar theories we have encountered. Since the generation of one thing, he says, necessarily entails the corruption of another, "when a more perfect form arrives, the prior form is corrupted; provided however that the succeeding form has everything that the first had, plus something more. And thus, through many generations and corruptions one arrives at the last substantial form, both in man and in other animals. ... Thus therefore it must be said that the intellective soul is created by God at the end of human generation, and this soul is at once both sensitive and nutritive, the preexisting forms having been corrupted."[32]

Thomas then has made the soul the form of matter, acting largely through material organs. But it was necessary for him to save the non-materiality of the intellect, for unless it were immaterial it would not be able to abstract species from the phantasms. He solved this by holding that although the soul is united essentially to the body, it has powers that flow from its own essence. Many of its powers complete their actions by means of the body (as in sensation), but it is not necessary that all of them should. Understanding is such a power flowing from the essence of the soul that completes its operation without any involvement in matter. In this way the soul can be the form of

---

[31] *In De anima comm.*, n. 220.

[32] "Et ideo dicendum est quod, cum generatio unius semper sit corruptio alterius, necesse est dicere quod tam in homine quam in animalibus aliis, quando perfectior forma advenit, fit corruptio prioris: ita tamen quod sequens forma habet quidquid habebat prima et amplius. Et sic per multas generationes et corruptiones pervenitur ad ultimam formam substantialem, tam in homine quam in animalibus aliis. ... Sic igitur dicendum est quod anima intellectiva creatur a Deo in fine generationis humanae, quare simul est et sensitiva et nutritiva, corruptis formis praeexistentibus." *Summa theologiae* 1, q. 118, art. 2, ad 2.

matter but still have an operation, understanding, that is unmixed with matter. This was a point that Siger would never concede to him.

Many of the elements of Thomas's solution had been proposed by his predecessors. The intimate connection between the body, as matter, and the soul, as form, was a prominent feature of the thought of Peter of Spain, and both Peter and Albert the Great had declined to consider the soul a complete *hoc aliquid* and had emphasized the soul's natural state of union with the body, without which it was incomplete. The doctrine of the progression of forms had an even longer history. It is based ultimately on Aristotle's discussion of the development of the embryo in *De generatione animalium* 2, 3. Philip the Chancellor had used almost the same words in describing the requirements: the succeeding form must possess everything that the prior one possessed plus something more, although in Philip's thought the earlier forms continued to exist; and William of Auvergne's example of the lower forms being absorbed by the rational soul implies a similar progression. Peter of Spain had taught that the rational soul was the only substantial form of the body and tried to work out an explanation of how it could be the form of matter while still retaining some power that flowed from its essence as a spiritual being and was not exercised through a bodily organ. John of La Rochelle too tried to depict it as the form of matter but not a material form. And many authors had taken from Boethius the doctrine that the soul, like any spiritual substance, was composed of *quod est* and *quo est* (for example, Philip the Chancellor, John of La Rochelle, Anonymous Admontensis, and Albert the Great), but they had meant a variety of different things by this formula. The most original alteration Thomas made in the thought of his predecessors was his reinterpretation of this formula, and his understanding of what the term 'substantial form' meant. By making prime matter, rather than the body, the subject of man's substantial form, and by positing the successive generation and corruption of forms in the embryo prior to the infusion of the rational soul, he solved the vexing problem of how the rational soul was related to the vegetative and sensitive souls.

Although Aquinas's solution of the problem of the soul was far superior to those of any of his contemporaries, and many modern scholars tacitly assume that it was a flawless solution, it did not impress his contemporaries the same way. The theologians were uneasy with his making the soul less than a *hoc aliquid*, with his assertion that it was individuated *ex corpore*, and especially with his understanding of the unity of substantial form. Many previous authors had conceded that there was only one substantial form of man and that this was the rational soul, but they considered the preformed body, a separate substance, as that of which the soul was the form; and they most often considered the soul to be compounded of vegetative, sensitive, and rational constituents, differing in origin and fate. But to Aquinas this meant that it was the rational soul that gave being and life to prime matter, constituting it as a living human being. The artists were particularly critical of his attempt to make the rational soul the form of matter—for it is not immediately apparent how it can be such and still not be a material form—and of his doctrine of the progression of forms. Aquinas's teaching on the soul did not carry all before it; it did however become the focus of most subsequent discussions of the soul, usually as the target of opposing views, nearly on a par with Averroes's doctrine of monopsychism, which was just beginning to be understood.

CHAPTER SIX

THE BEGINNING OF THE MONOPSYCHISM CONTROVERSY

By far the most difficult problem arising from Aristotle's *De anima* was the nature of the intellect and its relation to the body of the individual human being. There are many ambiguities and inconsistencies in Aristotle's account, to which we called attention in chapter 1, and indeed there seems to be no way to reconcile the various statements he made about it. A brilliant attempt to make Aristotle's teaching on the intellect consistent with his metaphysical principles and with what he said elsewhere about the rational soul was made by the Spanish Muslim Averroes in book 3 of his Great Commentary on *De anima*. Averroes himself had struggled with this question for many years, and his final view, which we shall examine here, represents a complete reversal of his earlier opinion as contained in his *Epitome* of *De anima*.[1] But brilliant as this attempt was in bringing consistency to Aristotle's teaching, this explanation raised insuperable difficulties for both the Muslim and Christian religions, as well as psychological problems in accounting for our experience of understanding.

AVERROES. At the beginning of book 3 of his Great Commentary on *De anima* Averroes had developed his famous interpretation of the human intellect. He began by pointing out that Aristotle had said that the intellect must be unmixed so that it might grasp and receive all things, for if it were mixed, then it would be a body or a power in a body; and if it were either of these, it would have its own form, a fact that would prevent its reception of the forms of other things. Although the intellect itself is not transmutable, it has something passive in it, inasmuch as it is moved by the 'forms in act' that it receives, to which it is in potency. But since it strips these forms of their material and particular determinations as it receives them, it must also be

---

[1]   See the excellent treatment of this development in Herbert A. Davidson, *Alfarabi, Avicenna, and Averroes on Intellect* (New York/Oxford, 1992), pp. 258-98. It is Davidson's opinion that Averroes seriously misread Aristotle on this point (pp. 298, 356), and as evidence of the still present confusion concerning Aristotle's true doctrine he cites E. Zeller, *Die Philosophie der Griechen* 2.2 (4th ed., Leipzig, 1921), pp. 573-75 and F. Nuyens, *L'évolution de la psychologie d'Aristote* (Louvain, 1948), pp. 303-04, 308, 311, who considered the intellect to be transcendent, J. Rist, "Notes on Aristotle *De anima* 3, 5," in *Essays in Ancient Greek Philosophy*, ed. J. Anton (Albany, 1972), pp. 506-07, who considered it to be a part of the individual soul, and W. D. Ross, who in his edition of Aristotle's works (5th ed., London, 1949), p. 153 considered it to be transcendent but had changed his mind by the time he edited the *De anima* (Oxford, 1961), pp. 45-47.

active. And therefore Aristotle says that it is necessary to posit a power of acting and a power of being acted upon in the rational soul, and he clearly says that both these parts are ungenerated and incorruptible.

But even though the intellect is passive in some respects, it is nevertheless completely unmixed with matter and has nothing of the nature of material forms. If it did, it would not be able to receive such forms, for thus it would receive itself, and the mover and thing moved would be the same. The material intellect (the name Averroes uses for the possible intellect) is potentially all meanings of all material forms, and it is not any being actually before it understands something. Its receptivity differs from the potency of prime matter (which is similarly in potency to all forms), since the intellect is in potency to universal forms, which it distinguishes and grasps, whereas prime matter is in potency to individual forms, and it does not distinguish or grasp them.

Then Averroes raises a question that he says is of very great difficulty: If the possible intellect is the first perfection of man, as it is according to Aristotle's definition of the soul, and the speculative intellect is his last perfection, but man is generable and corruptible and is numerically one being through his last perfection by the intellect, it is necessary that it (i.e., the speculative intellect) be his first perfection, since through the first perfection of the intellect, I am other than you and you are other than me; and thus man would not be generable and corruptible in the respect in which he is man, but in the respect in which he is an animal. It is thought that, in the same way that it is necessary that, if the first perfection were a *hoc aliquid* and numerable by the number of individuals in order that the last perfection might be of this sort, thus also the contrary is necessary, namely that if the last perfection is numbered by the number of individual men, then the first perfection should also be of this sort.

Averroes then treats a number of impossible consequences of this, providing in the process many of the arguments that the Latins would use against him, investigating the views of Alexander of Aphrodisias, Abubacer, Avempace, and Theophrastus, and finally says: "Let us now return to our own views. ... And since there are all these different opinions, it seems to me that I should expound this matter as it seems to me. And if what appears to me is not complete, it will be the beginning of a full account. And thus I ask the brothers who see this work to write down their doubts, and perhaps in this way the truth will be found, if I have not yet found it. And if I have found it, as I think, then it will be made more plain by these questions."[2] This seems

---

[2] "Revertamus igitur ad nostrum. ... Et cum ista sint, ideo visum est michi scribere quod videtur michi in hoc. Et si hoc quod apparet michi non fuerit completum, erit principium complementi. Et tunc rogo fratres videntes hoc scriptum scribere suas dubitationes, et forte per illud invenietur verum in hoc, si nondum inveni. Et si inveni, ut

clearly to indicate that he is stepping out of his role as commentator on Aristotle's text and undertaking an original solution, consistent with Aristotelian principles, to a question that the Philosopher had left unsolved.

He first treats the question of how speculative intellects are generable and corruptible, while the agent and possible intellects are eternal. But his second question is of much greater importance and interest: How the material (i.e., possible) intellect is one in number for all individual men, neither generable nor corruptible, and the things actually understood by it are numbered by the number of individual men, generable and corruptible because of the generation and corruption of individuals. "Indeed," he says, "this question is of the greatest difficulty and contains the greatest ambiguity."[3]

He goes on to show that regardless of which of the possibilities we choose, we come to an impossible conclusion. If we posit that the material (i.e., possible) intellect is numbered according to the numeration of individual men, it would be a *hoc aliquid*—either a body or a power in a body. But then it would be the understood material form in potency, and therefore a subject that moves the possible intellect, which would thereby receive itself; and this is impossible. And even if we should concede that it does receive itself, then there would be no difference between the form outside the soul and the form in the soul—that is, they would both be material forms.

And if we posit that it is not numbered by the numeration of individuals, then it would happen that its relationship to all the individuals of which it is the last perfection would be the same, and therefore it would be necessary that if some of those individuals should acquire some understood thing, that that thing should be acquired by all of them. Since, if the conjunction of those individuals with what is known results from the conjunction of the material intellect with them, just as is the case with the senses, it is necessary that if you aquire some knowledge, I shall also acquire it; which is impossible. The intellect then must be a separated substance, and it must therefore be unique. And therefore one must hold the opinion that, if there are some beings through whose first perfection there is a substance separated from their substance—as they think is true of the heavenly bodies—that it is impossible that there be found more than one individual of one such species, for if there were more than one, it would be superfluous, just as it would be superfluous to have two pilots for one ship.

---

fingo, tunc declarabitur per illas questiones." *Commentarium Magnum in Aristotelis De Anima Liber Tertium*, ed. Crawford, pp. 398-99.

[3]   "[H]ec quidem questio valde est difficilis, et maximam habet ambiguitatem." *Op. cit.*, p. 402.

But if this is the case, how does it happen that each individual seems to think his own thoughts? We may say therefore that it is clear that man only understands in act because of the conjunction with him in act of the thing understood. And it is also clear that matter and form are reciprocally joined so that the thing made from them is one thing, and especially the material intellect and the understood notion in act, for what is composed of them is not some third thing other than them, as is the case with other things composed of matter and form. Therefore, it is impossible that the conjunction of the understood thing with man should be anything else but the conjunction of one of these two parts with him, that is, of the part that is like matter, or of the part that is like form.

And since we have established that it is impossible that the thing understood is coupled with each and every man and numbered by their numeration through the part that is like matter, i.e., the material intellect, it remains that the conjunction of the thing understood with us men be through the conjunction of the intellected meanings with us, that is, of the part of them that is in us in some manner like form.

And therefore one must hold the opinion (which has by now become apparent from Aristotle's words) that in the soul there are two parts of the intellect, of which one is the recipient, the other the agent, and it is the latter that makes the meanings that are in the imaginative power, and that these two parts are neither generable nor corruptible.

The material and agent intellects seem in one way to be two things, and in another way one thing. They are two through the diversity of their action, for the action af the agent intellect is to generate, and of the material intellect to be informed. But they are one thing because the material intellect is perfected by the agent and knows itself. And in the manner that we say that the intellect is joined with us, there are apparent in it two powers, one active and one passive.

Averroes then holds that there is only one possible and agent intellect for the human race. It is devoid of matter and therefore cannot be multiplied. The phantasms provided by the senses are received by the possible intellect. The possible intellect is then acted upon by the agent intellect, which strips them of their particular and material conditions, thus providing universal forms and brings about actual understanding from potential understanding. But the possible intellect, in itself, contains nothing, and everything in it is provided by the phantasms of individual men. It is the locus of understanding, for it is upon it that the agent intellect acts. Because the possible intellect is by nature continuous with the phantasms, it can communicate this understanding to each individual respecting only his phantasms, for it is a process or activity, not, prior to receiving forms from the phantasms, a repository of known objects. Hence the knowledge of each man is his own, and it is this individual body of understanding that constitutes the speculative intellect, which, since it depends on the sense knowledge provided by the body, is perishable. It is however the speculative intellect that constitutes the specific difference that makes one individual distinct from another, thus by implication denying personal immortality. It is hardly surprising that Christian theologians were vehement in their denunciation of Averroes's views. It is a good deal more surprising that it took them so long to realize what he was saying.

ALBERT THE GREAT, *DE UNITATE INTELLECTUS.* For approximately the first twenty years that Averroes's *Commentarium magnum* on *De anima* had been available to the Latins, it seems that no one realized that he had taught the possible, as well as the agent, intellect to be unique for the human race and not a part of the rational soul of each individual. Such disagreement as there was had to do with the agent intellect. But by the middle 1250s his doctrine came to be generally recognized. I do not know who was the first to realize this, but both Thomas Aquinas and Bonaventure were well aware of it when they composed their commentaries on the *Sentences* during the first half of the 1250s, and by 1256 Albert the Great, who had by now corrected his earlier misunderstanding of Averroes, was asked by pope Alexander IV to write a refutation of Averroes's position that there is a single agent and possible intellect for the entire human race. The result was a succinct account of the pro-Averroistic arguments, of the anti-Averroistic arguments, and a brief but clear essay summing up what Albert considered to be the correct position. There is no hint in Albert's treatise of anyone's seriously maintaining Averroes's view. It is a businesslike, non-polemical rehearsal of the arguments for and against, with a conclusion based on the arguments. Albert found Averroes's position to be heretical, to be sure, as well as unsupportable on purely philosophical grounds. He exhibits no alarm or undue concern but simply sets about refuting the erroneous contention that there is a single intellect for all humans. On the description of the process of intellection, however, he found Averroes to be superior to the other philosophers and made the Muslim's position his own.[4]

Albert says in chapter 1 that he will leave aside what the Christian faith teaches on the subject and confine himself to those things that can be demonstrated by syllogisms. In short, this is to be a philosophical, not a theological, work. At the beginning of chapter 3, he warns that this is a very difficult question and could be understood only by those who have been nourished by philosophy. He lists the questions concerning the soul on which the authors of the philosophical tradition are in doubt but says that he will

---

[4] See for example his *De anima*, 3, 3, 11, ed. S. C. A. Borgnet, *B. Alberti Magni, Ratisponensis Episcopi, Ordinis Praedicatorum, Opera Omnia* (38 vols., Paris:Vives, 1890-1899), V, p. 385.

confine himself to the question of whether there is a single intellect for the human race. Then, after a rehearsal of some of the more important positions of the Peripatetics and Arabs, he presents thirty arguments in favor of the uniqueness of the intellect, taken for the most part from Averroes's *Commentarium magnum in De anima* and authors quoted there. This is followed by thirty-six arguments for the other side, the most important of which are the eighth, twelfth, twenty-third, twenty-fourth, and thirty-fourth.

The eighth argument is based upon the proper nature of the soul, and Albert considers it to be powerful (*fortis*): In every nature there is something active and something passive. In the soul, the passive element is the speculative intellect. It is a subject and is generated not by the action and generation of nature, but by the action and generation of the soul, in an absolute sense (*simpliciter*), since no principle of nature acts on, or is acted on, by such a made thing. Therefore the very nature of the soul must be something acting *animaliter* and undergoing *animaliter*. But since there are as many speculative intellects as there are human beings, the same is true of the possible intellect. Therefore there will be as many possible intellects remaining after the death of the bodies as there are [dead] men of which they are the possible intellects.[5]

In the twelfth argument, Albert uses Averroes's explanation of Aristotle's argument near the end of *De anima* to show that there is one soul (not three) for each human body: There is one substance of the soul. For just as the body is one thing, whose parts are members which are called organs, thus the soul, which is the perfection of the body, is one substance whose potestative parts are powers, which are called potencies of the soul. But if this is true, then the vegetative and sensitive and rational in man are one substance. But something that is one and the same in substance cannot be partly numbered according to the number of those things of which it is the substance (i.e., the vegetative and sensitive) and partly one thing existing one in number and common to all men (i.e., the rational). Therefore, since the vegetative and sensitive are numbered according to the number of men in which they are, so will be the rational. At the end of this argument, Albert departs from his discussion of monopsychism to dismiss the view of certain Latins, saying that the contention that there are three substances in man is completely ridiculous and is not a doctrine of the Peripatetics but of "certain Latins who are ignorant of the nature of the soul." It is, he says, "a most foul error to say that there are several substances in one subject, since those substances can only be forms. But there can be many powers of one substance."[6]

---

[5] *De unitate intellectus, ed. cit.,* IX, p. 454a.

[6] *Ibid.,* p. 155a-b.

The twenty-third argument is of particular interest: If man is distinguished from other animals by intellect alone, and "a man does not generate unless he has a sensible soul, it follows that man does not generate man, but an incomplete animal does, because the generation of man does not include the highest (*ultimam*) form, nor is that highest form any part of man's being. All of which is completely absurd."[7]

In the twenty-fourth argument, Albert gets to the heart of the difference between the Christian and Averroistic doctrines of the rational soul, that is, whether or not it is a 'separate substance' in the Aristotelian sense. If we assume that it is, he says,

> then it will always follow that it would be separate according to substance and being, because it is neither the entelechy of an organic body having the capacity for life, nor is it the principle of this life, because a separated substance does not touch, and if it does not touch, it does not act, nor does it perform operations, nor is it the cause of these operations. Therefore it is not a soul. But everyone admits that it is a soul. But if someone should say that an intellectual substance is in everyone in such a manner that it is in no one *per se*; this is ridiculous, because nothing of natural things is in all things except that which is in each of them.[8]

The thirty-fourth argument follows up his point that the soul cannot be an Aristotelian separated substance: The simultaneous destruction of all men, which is a possibility even though Averroes denies it, would leave a separated substance that is idle and vain, moving nothing and doing nothing. "This is false and impossible, and so the hypothesis [that the soul is a separated substance] is also."[9]

---

[7] "Si enim homo est solus intellectus, et homo non generat nisi habens animam sensibilem, et nihil de intellectu, sequitur quod homo non generat hominem, sed animal imperfectus: quia generatio hominis non includit formam ultimam, nec ultima forma secundum esse est aliquod de homine: quae omnia valde sunt absurda." *De unitate intellectus, ed. cit.*, p. 459a-b.

[8] "... semper hoc sequitur, quod est separatum secundum substantiam et esse: quia nec est indelechia corporis organici potentia vitam habentis, nec est causa et principium hujus vitae: quia separatum non tangit: et si non tangit, non agit, neque operatur, <nec> est causa operationis. Igitur non est anima. Quod autem est, omnes confitentur. Si autem dicat aliquis, quod substantia intellectualis ita est in omnibus, quod in nullo per se: hoc ridiculum est, quia de rebus naturalibus nihil est in omnibus, nisi quod est in quocunque eorum." *De unitate intellectus, ed. cit.*, p. 459b.

[9] *De unitate intellectus, ed. cit.*, pp. 461b-462a.

WILLIAM OF BAGLIONE. In the present state of our knowledge, it appears that
the leader in mounting an emotional opposition both to the monopsychism of
Averroes and the teaching of Thomas Aquinas was the Franciscan master
William of Baglione, who was regent in theology at Paris in 1266-67. His
basic positions owed much to the thought of his confrère, and by now
Minister General of the Minorites, Bonaventure, but he was also a powerful
thinker in his own right and did not simply parrot Bonaventure's doctrine.
The influence of Grosseteste is also noticeable, especially in William's
emphasis on the soul as the image of God and his use of many *passus* of
Augustine that Grosseteste had used but which were not generally cited in
disputations on the soul. William's primary concern was monopsychism, and
he was much concerned by the consequences of Averroes's teaching that if
there were one intellect for all men, it would not strictly speaking be true that
a man understands, a position very likely being maintained at this time by
Boethius of Dacia.

So exercized was William by the doctrine of monopsychism that he inserted
mention of it into one of his disputed questions on the eternity of the world.
He mentions there several opinions which he holds to be erroneous: 1-
Algazel's contention that an actual infinity of separated souls could exist,
since they are not causally related; 2- Averroes's teaching that there is only
one soul for the human race, a particularly dangerous error, which was the
basis of the subsequent ones; 3- that the soul is individuated by the body; 4-
that the intellect, or intellective soul, is not a *hoc aliquid*, or individual thing;
and 5- that the soul is *ex traduce*, that is, it is generated by biological means
by the parents. This last error, he says, "is inconsistent with the dignity of the
image [of God]."[10]

William attributes the origin of these errors to Averroes but complains that
"today there are blind leaders of the blind" who are teaching them. He
mentions no names, but his words seem clearly to indicate Aquinas as well as
certain masters in the arts faculty. And he promises to impugn the error of the
Commentator concerning the unity of the intellect in a special question below
("Ille error Commentatoris de unitate intellectus quaestione speciali inferius
impugnatur.")[11]

---

[10]   William of Baglione, "Utrum mundus habuerit suae durationis initium vel sit
ponere mundum ab aeterno," ed. Ignatius Brady, O.F.M., "The Questions of Master William
of Baglione, O.F.M., *De aeternitate mundi* (Paris, 1266-1267)," Part II, *Antonianum* 47
(1972), 576-616, on p. 604.

[11]   *Loc. cit.*

The specific question which seems to correspond to this promise is found in Vatican MS Palat. lat. 612, folios 158va- 159rb, "Utrum in omnibus hominibus sit intellectus unus numero," and has been edited by Fr. Ignatius Brady[12] along with two other questions of William's on the soul, "Si illud quod constituit hominem in esse specifico operatione naturae educatur de potentia materiae vel per infusionem animae rationalis," and a second, more polished, version of the same question contained in Florence, Biblioteca nazionale, MS conv. soppr. B.6.912, folios 22rb-23va, as well as one version of the question "Utrum anima rationalis ex natura sua sit hoc aliquid," from folios 54bis rb- 55rb of the Florence manuscript. The first of these is concerned with the fundamental question of the uniqueness of the intellect. The next two treat the traditional question (number five in the list given above), which had so troubled Augustine and Gregory the Great, whether the soul is *ex traduce*. The last concerns the fourth of the errors William had noted in his question on the eternity of the world, whether the soul is a *hoc aliquid*.

William objects to any form of the compound soul theory and adopts an explanation of the embryo's development prior to the infusion of the rational soul very similar to that of Roland of Cremona, Peter of Spain, and Albert the Great. Although some people say that the vegetative and sensitive powers that preceded the infusion of the rational soul are united with it to make one thing, and that when a man dies the rational soul is separated from the body while the lower parts perish with the body, William says that it is otherwise: The spirit and vigor and natural heat are deposited with the semen, and they are warmed in the mother's womb until they bring the body to the point of development that, by God's command, the rational soul is infused, carrying with it its own sensitive and vegetative powers.[13]

He insists that the soul is a *hoc aliquid*, consisting of spiritual matter and form, but he is also much concerned to emphasize the unity of the human individual and the soul as the form and perfection of man. He does not hold that the body has its own form but teaches that the rational soul is the form of the body and consequently of man. To the objection that if the rational soul performed all the vital functions, then after its separation from the body it would be idle, he replies that while the soul is united to the body it is suited to move the body because it perfects it. But when it is separated from the body

---

[12]   Ignatius Brady, O.F.M., "Background of the Condemnation of 1270: Master William of Baglione, O.F.M.," *Franciscan Studies* 30 (1970), 5-48.

[13]   William of Baglione, "Si illud quod constituit hominem in esse specifico operatione naturae educatur de potentia materiae aut per infusionem animae rationalis," ed. Brady, *art. cit.*, p. 17.

it is not idle, because it is not then perfecting it. A thing is only called idle when it is not doing what it is supposed to do. And furthermore, the soul performs other operations than moving the body, such as sensing and understanding, and it continues to do this after its separation.[14]

In explaining why Averroes came to the conclusion he did on the matter of the intellect, William provides a full and clear summary of the Muslim's reasoning. But this position, says William, implies that a man does not understand. According to Averroes's account, a man understands through the conjunction of our (speculative) intellect with the illuminated forms, and not from the conjunction of the man who is understanding with the possible intellect; and in this way he can hold that the intellect is always understanding, because it is always bringing about understanding in someone.

And he says that the thing that is understood, insofar as it concerns our intellect, is not in the forms, and therefore what I understand, someone else does not, and the other way around.

But William considers this argument to be insufficient. Universal form, or universal, he says, may be understood in two ways: it can refer to the property by which a thing is called one as opposed to many; or it can refer to the fact that a thing can be predicated of several things. And when one says that the intellect does not understand those particular and universal forms, 'universal' is understood in the first way, rather than the second; "that is, it abstracts the quiddity and purity of a thing from its material appendages and from here and now. Whence it only understands things that it has abstracted from the senses and through the senses." But this is false, William claims. Rather, it also understands what it has received from a higher influence, and it understands itself and its powers, and an angel, and faith, and it does this through its substance. "And thus it not only understands a universal but also a particular as a singular and that it is not predicated of several things. Nevertheless that particular assumes the form of a universal insofar as it is abstacted from the appendages of matter and from the here and now."[15] So it is false to say that we only understand through the conjunction of the species provided by the imagination and phantasms, because the intellect does not understand the latter insofar as they are such. Rather, understanding belongs to the possible intellect as a result of its conjunction with the species provided by the

---

[14] William of Baglione, "Utrum in omnibus hominibus sit intellectus unus nume-ro," ed. Brady, p. 43.

[15] "[E]t sic non tantum intelligit universale et particulare ut singulare et quod non praedicatur de pluribus. Tamen illud particulare induit formam universalis in quantum abstrahitur ab appendiciis materiae et ab hic et nunc." *Op. cit.*, p. 39.

phantasms. "And therefore it must be said that the intellect is diverse in diverse men."

In his expanded Response to the question "Utrum anima ex sua natura sit hoc aliquid," William attempts to solve one of the fundamental questions involved in the soul-body relationship, namely how the soul can be a distinct individual substance and still function as the substantial form of the body. "There can be no doubt," he says, "that the rational soul is, by its own nature, a natural being and a *hoc aliquid* and is by its own natural inclination the true and first perfection of man, so that it is the true perfection of the human body according to the definition of the Philosopher in book 2 of *De anima* that the soul is the perfection of a natural organic body having the potency for life."[16]

He then points out that there are four conditions that lead us to call something a *hoc aliquid*. First is designated or 'stamped' being (*esse signatum*), as is indicated by the demonstrative pronoun 'this' (*hoc*). Second is its own being (*esse proprium*). Third is permanent being (*esse permanens*), such that it can subsist by itself without something else. And fourth is independent being (*esse independens*). These conditions exclude such things as universals, accidents (such as whiteness or roundness), and things such as matter and form, which exist only as constituents of a *hoc aliquid*. The rational soul satisfies three of these conditions. It has designated being. It has its own being, because it is composed of matter and form corresponding to it as a spiritual being. It has permanent being fixed in itself, because it is incorruptible and immortal when it is separated from the body. But it does not have the fourth, namely independent being, but is rather the true perfection of the human body, with which it makes one thing. It is therefore a form in relation to the body, but neither in the way that God is the form of all things by forming them and giving them being, nor in the way the souls of beasts are the forms of their bodies, acting solely through bodily organs. The human soul performs all the vegetative and sensitive functions that the souls of beasts perform using bodily organs, but its highest power, the rational or intellective, is exercized without the use of bodily organs. Through the intellective act, the soul is reflected above itself and in the process necessarily abandons whatever is corporeal. "If the dignity of the soul is necessarily led to God by the act of its own powers, it is clear that it is a *hoc aliquid*. If it administers and rules its own body as a perfectible thing with which it makes

---

16 "Absque dubitatione tenendum est ... quod anima rationalis ex sua natura, hoc est ex suo et sibi naturale esse, est hoc aliquid, et ex sua et sibi naturali inclinatione est vera et prima perfectio hominis, ita quod est vera perfectio humani corporis secundum illam diffinitionem animae quam ponit Philosophus in II *De anima*, quod scilicet 'anima est perfectio corporis organici, physici, potentia vitam habentis." William of Baglione, "Utrum anima ex sua natura sit hoc aliquid," ed. Ignatius Brady, O.F.M., "Background to the Condemnation of 1270," p. 28.

one being, it is clear that it is the true and first perfection of the human body
as it is susceptible of life and sense and motion. Therefore, Aristotle's
definition applies to it properly *(proprie)*, not equivocally as it is understood
by the nonsense of the Commentator."[17]

William then launches an attack against Averroes and those who follow him
in this matter. In his view, if one denies that the soul is a *hoc aliquid* and that
it is composed of spiritual matter and form, it would be necessary to hold that
it is individuated by the body and so would perish with the body. And in his
condemnation of his opponents he mixes together purely Averroistic positions
with several of Aquinas's doctrines (not always properly understood). "Those
who posit the contrary," he says,

> namely that the rational soul is not a *hoc aliquid* and that it
> is not the first perfection of the human body and thus not
> the first perfection of man, share in that most pernicious
> error of the Commentator concerning the unity of the
> intellect. This error is founded on these two points, as is
> completely clear to anyone who inspects and reads the
> words of Averroes in the passage from which these errors
> flow, although some people wish to defend one without the
> other. Those who hold that the rational soul is not a *hoc
> aliquid* but is individuated by the body come close to
> agreeing with Averroes, because the soul's being
> individuated by the body can only be understood if the soul
> should receive from the body or in the body that which is
> the cause of all individuation in the completed being. But
> this is matter. Therefore, the soul receives matter either
> from the body or in the body as part of itself. This is totally
> absurd, especially since those who follow Averroes, when
> they dispute about the simplicity of the soul, completely
> deny that it has matter.[18]

---

[17] "Si igitur animae dignitas hoc necessario habet quod fertur in Deum per actus
suarum virium, clarum est quod vere est hoc aliquid. Si administrat corpus suum et regit sicut
perfectibile cum quo facit unum esse, manifestum est quod [est] vera et prima perfectio
humani corporis ut est susceptibile vitae et sensus et motus; [et quod] vere et proprie
communicat ei diffinitio illa animae quam assignat Philosophus in secundo *De anima*, non
aequivoce secundum deliramentum Commentatoris." *Ibid.*, p. 33.

[18] "Contrarium autem ponentes, quod scilicet anima rationalis non est hoc aliquid, et
quod non est prima perfectio humani corporis, et sic non est prima perfectio hominis, fovant
illi perniciosissimo errori Commentatoris de unitate intellectus. Ista enim duo principalia sunt
fundamentum illius erroris, sicut manifestissimum est inspicienti et legenti verba Averrois in
illa passu. Propter quod isti errores consequuntur se, quamvis aliqui velint unum defendere
sine alio. Qui autem ponunt quod anima rationalis non est ex se hoc aliquid, sed per corpus
individuatur, propinqui sunt illi favori, quia animam per corpus   individuari non potest

According to the Philosopher, William continues, and according to the common opinion of all those who are acquainted with philosophy, since 'being' and 'one' are convertible, a thing receives its essential unity and its being from the same agency, and it must also receive its individuation from the same thing as that from which it receives its unity. Therefore, if one holds that the soul is individuated by the body, it follows that it receives its being from the body. And this opens the way to manifold error, for if it is thus, then the soul perishes with the body, and all the things that scripture contains concerning the dignity of the rational soul, human reparation and justification, as well as the degrees of rewards and punishments, will be voided. Furthermore, if the soul is individuated by the body, when it is separated, it would not be a *hoc aliquid* or an individual. To this, he says, some people respond by citing the example of Avicenna (cf. *De anima* 5, 3) of a coagulated liquid that retains the form of its container. But, he says, this example is not pertinent here, for the way in which the soul is in the body is not even remotely similar to the way a liquid is in its container. Fr. Brady has pointed out that the example is that of Aquinas, *Comm. in Sent.* I, d. 8, q. 5, art. 2, ad 6.[19]

William was among those who emphasized the natural unity of body and soul. The soul does not possess independent being but is suited to be the substantial form of the human body, and this is part of its nature, although it is capable of existence apart from the body. Although its essential function is to animate the body, it is not related to the body only as a mover. It is "the perfection of the human body and, as a consequence, the perfection of man. Since it gives being to that which is perfectible by it, constituting one being with it, therefore it is truly and formally a perfection,"[20] having an act in the body (i.e., pumping blood, digesting food); an act through the body (i.e., voluntary actions); and an act above the body (i.e., understanding).

---

intelligi nisi anima recipiat a corpore vel in corpore illud quod secundum ipsos causa est omnis individuationis in ente completo. Hoc autem est materia. Ergo materiam recipit anima vel a corpore vel in corpore tamquam partem sui. Hoc autem absurdissimum est dicere, ut manifeste sunt sibi ipsis contrarii, quia cum disputant de simplicitate animae, omnino negant eam habere materiam." *Loc. cit.*

[19] Brady, *art. cit.*, p. 34, n. 28.

[20] "[Ex his patet quo sensu dicitur] anima rationalis perfectio humani corporis, et ex consequenti perfectio hominis. Quoniam scilicet suo perfectibili dat esse, unum esse cum eo constituens, ideo est perfectio vere et formaliter." *Ibid.*, p. 31. The remainder of the sentence is a conflation of what William says here and on pp. 29-30.

Although Aquinas's doctrine was clearly the object of a portion of William's attack, the two masters have much in common. Both men are very concerned to account for the unity of the individual human being while still allowing for the soul to survive the body's death. Both think highly of Aristotle and accept his definition of the soul, while condemning Averroes as the perverter of Aristotle. Both insist that the soul is the true substantial form of the body according to its first act, although they meant quite different things by the term 'form.' But William also had some basic disagreements with Thomas. He adopted the doctrine of Avicebron that the soul was composed of spiritual matter and form to enable him to maintain that the soul possesses *esse proprium* and thus is a *hoc aliquid,* as well as to account for its individuation. William also denied Thomas's view of the soul as not being a *hoc aliquid,* although he conceded that it lacked one characteristic of a *hoc aliquid,* namely independent being. And William could not see that the soul could be individuated if it did not contain matter; if it were individuated by the body, then when it was separated from the body it would cease to be individual.[21] This is an interesting point, and it was also used by the 'Averroists' to argue on behalf of a unique possible intellect, although, unlike William, they denied that the intellect contained matter.

William is the first master of whom we have knowledge to become seriously concerned about the nature of the teaching by some members of the Parisian arts and theology faculties. This concern is evidenced by the fact that he composed three versions of his question on the eternity of the world and disputed three separate questions on the soul related to the doctrines that he considered erroneous, two of which exist in two versions. On this basis we may fairly consider him a master of considerable ability and importance. It is clear enough that he was a leader in the movement against certain positions, namely the uniqueness of the intellect and the eternity of the world, which were derived from Averroes's commentaries on Aristotle as well as certain doctrines of Aquinas that he considered dangerous. It is equally clear that he does not display any hostility to Aristotle; indeed, he accepts his definition of the soul, as far as it goes, and blames its perversion on Averroes.

JOHN PECHAM. John Pecham was regent master in theology at Paris from 1269 to 1271, then lectured for several years at the Franciscan convent at Oxford, and in 1277 was appointed lecturer at the papal curia. In 1278 he succeeded Robert Kilwardby as archbishop of Canterbury. In this office he wrote a letter to the bishop of Lincoln (June 1, 1285) deploring the dangerous novelties which had invaded theological teaching during the preceding twenty

---

[21] *Loc. cit.*

years and the abandonment of the saints of old, especially Augustine,[22] and in 1284 he renewed Kilwardby's condemnation of 1277 (although no one could find a copy of the condemned articles, so he was not sure what he was condemning),[23] to which he added specifically the doctrine of the unity of substantial form. But during his academic career at least he was certainly a middle of the road theologian and was far from being an arch-conservative. He himself was very much involved in the 'novelties' of which he complained after he became archbishop; he made extensive use of Avicenna, Gundissalinus, Aristotle, Averroes, and especially Avicebron, not only quotations but basic positions.[24]

Pecham's doctrine of the soul is contained in a series of disputed questions,[25] which date from his Parisian regency (1269-71), and his treatise *De anima*,[26] which was probably composed between 1277 and 1279, when he was a lecturer at the papal university. It has much in common with that of William of Auvergne. He takes very seriously the question of the soul's being *ex traduce*, especially because both Augustine and Gregory the Great had expressed doubts, but concluded that it was not. He parts company with William of Baglione on the relation of the vegetative and sensitive souls to the rational, opting for the compound soul theory. The soul of the embryo is passed down from the parents to the child, but when the rational soul is created and immediately infused into the body, it perfects and completes, rather than corrupts, the earlier soul. Therefore, there are not two souls in man, but only one, the rational soul: "The soul which is generated and the

---

[22] C. T. Martin, ed., *Registrum epistolarum fratris Johannis Peckham*. Rolls Series (London, 1885), III, p. 901.

[23] See Leland E. Wilshire, "The Oxford Condemnations of 1277 and the Intellectual Life of the Thirteenth-Century Universities," in *Aspectus et Affectus*, pp. 113-24, on p. 116.

[24] This point is aggressively argued by James A. Weisheipl, O.P., "Albertus Magnus and Universal Hylomorphism: Avicebron," in *Albert the Great. Commemorative Essays*, ed. Francis J. Kovach and Robert W. Shahan (Norman, Oklahoma, 1980), pp. 239-60.

[25] P. Hieronymus Spettmann O.F.M., ed., *Johannis Pechami Quaestiones Tractantes de Anima*. BG-PM XIX, 5-6 (Münster, i.W., 1918).

[26] P. Gaudentius Melani, O.F.M., ed., *Tractatus de anima Ioannis Pecham*. Biblioteca di Studi Francescani 1 (Florence, 1948).

rational soul which is infused are not two souls, but one, just as man is one substance composed of body and soul."[27]

This rational soul is the substantial form of the body with respect to its first act, for it gives being to matter and therefore it is multiplied as matter is. But with respect to its second act, which is understanding, it is immaterial, "because such an act does not go forth from potency by means of a mediating organ, since it is reflected on itself."[28]

When Pecham says the soul is 'immaterial,' however, he means only that it is not involved in corporeal matter, for, like Bonaventure and William of Baglione, he bases his position on Avicebron's teaching that the soul is composed of matter and form, both spiritual, a view he attributes to Augustine and even Averroes. The rational soul does indeed contain matter, though spiritual rather than corporeal. Therefore it is free of transmutable matter, "but it has something similar to matter itself through which it is 'this,' through which there is a natural distinction in separated substances."[29]   (It is this that confirms his true source to be Avicebron, not Augustine, for Augustine's reason for positing something like spiritual matter was specifically to guarantee the mutability of spiritual substances.) Also, it is not necessary that the intellect be completely free of matter in order to be the perfection of corporeal matter; it is only necessary that it should not essentially depend on matter.

> Indeed, just as there is a spirit that is totally free of [corporeal] matter in being and acting, such as an angel, which is suited to acquire for itself every perfection from above; and just as there is a spirit completely conjoined [to matter] in being and acting, such as a generable and corruptible spirit, which is completely perfected from below; thus there is a mean, which does not depend in

---

[27] "Et quod generatum et anima rationalis, quae infunditur, non sunt duae animae sed una. Sicut homo est una substantia ex anima <et corpore> composita." *Quaestiones tractantes de anima*, Q. 1, D, ad 7, *ed. cit.*, p. 9.

[28] "Unde est immaterialis respectu actus secundi, qui est operatio, quae est intelligere, quia actus iste non exit a potentia mediante organo, cum in se reflectatur." *Ibid.*, Q. 4, ad 14, *ed. cit.*, p. 56.

[29] "Est ergo immaterialis per exclusionem naturae transmutabilis, sed habet aliquid simili ipsi materiae, per quam est hoc, per quam est distinctio naturalis in substantiis separatis." *Ibid.*, Q. 4, Resp., *ed. cit.*, p. 50.

being but is united in acting, so that it acquires its perfection partly from above and partly from below.[30]

Some people, he says on the authority of Boethius (*De duabus naturis* 6), deny that the rational soul is composed of matter and form, because a spiritual substance does not have matter but is composed of quiddity and being, *quo est* and *quod est*. Pecham, misattributing Gundissalinus's *De unitate et uno* to Boethius, claims that Boethius contradicts this position by asserting that the genus 'substance' is divided into corporeal and spiritual, predicated of them univocally. "And so it is clear that every substance that is in a genus is composed of matter and form. Therefore it must be conceded through the aforesaid authorities and arguments that the soul is composed of matter and form."[31]

In discussing the agent intellect, Pecham makes a distinction between the separate agent intellect "of which the Philosopher speaks," which is God, and the active power of the soul that brings about actual understanding.   In Question 6 he says:

> The agent intellect about which the Philosopher speaks is not truly a part of the soul, but is God, so I believe, who is the light of all minds, from whom is all understanding. For it is he alone to whom belong all those noble properties about which the Philosopher speaks—that he is unmixed, impassible, and always knowing all things, whose substance is his action.[32]

It is the same as the *lumen* of which Augustine speaks, the uncreated eternal light.

---

[30] "Immo sicut est spiritus omnino a materia absoluta in essendo et perficiendo, ut angeli, qui nati sunt omnem perfectionem sibi acquirere a superiori, et sicut est spiritus omnino coniunctus in essendo et perficiendo sive agendo, ut spiritus generabilis et corruptibilis, qui omnino perficitur ab inferiori, ita est medius, qui non dependet in essendo, unitur tamen perficiendo, ut et ipse partim a superiori, partim ab inferiori perfectionem acquirat." *Ibid.*, Q. 4, Resp., *ed. cit.*, p. 50.

[31] "... manifestum est omnem substantiam, quae est in genere, esse compositum ex materia et forma. Concedendum est igitur per a<u>ctoritates et rationes praedictas animam esse compositam ex materia et forma." *Ibid.*, Q. 25, Resp., *ed. cit.*, p. 187.

[32] " Intellectus siquidem agens, de quo Philosophus loquitur, non est usquequaque pars animae, sed Deus est, sicut credo, qui est lux omnium mentium, a quo est omne intelligere. Ipse enim solus est, cui conveniunt omnes proprietates illae nobiles, de quibus loquitur Philosophus. Quia est immixtus, impassibilis et semper omnia intelligens, cuius substantia est sua actio." *Ibid.*, Q. 6, Resp., *ed. cit.*, p. 73.

But there is also an active principle in the individual soul:

> The position of Avicenna is better than that of those who
> posit the agent intellect to be only a part of the soul.
> Nevertheless, there is something active in the rational soul,
> which I call its power, through which it is suited to
> transform itself into a similitude of all intelligible things.
> Therefore, this power, if it be called the agent intellect,
> which also has something of natural created light, differs
> essentially from the possible intellect as power (*vis*) from
> power, not as potency from potency. For I believe that
> there are various powers of the same potency, as in the
> same organ, the eye, brightness (*splendor*) and clarity
> (*perspicuitas*) differ. Just so, the higher and lower reasons
> [of Augustine] are called the same thing related in different
> ways. The material intellect embraces the higher and lower
> reasons, and the created agent intellect [of the individual
> soul] perfects them both. ... The agent intellect is coupled
> with higher and lower things only through the possible
> intellect, ... which is suited to be illuminated from above
> and from below.[33]

By the time of Pecham's regency in theology, the unrest over the teaching of
some of the arts faculty had been evident for several years. Pecham devoted
one *quaestio* to "Whether there is one intellect for all men." He begins his
Response by stating what he considered to be the true position:

> As the Catholic faith hands down and the opinions of the
> saints agree in attesting, there is one rational soul for each
> individual human being, which is infused into the body as
> soon as it is created. Nor has any saint ever doubted this
> opinion, although some have been unsure about the origin
> of the soul, that is, whether the souls of each person are
> created from nothing or whether they are passed down from

---

[33] "Et pro tanto melius posuit Avicenna — qui posuit intellectus agentem esse
intelligentiam separatam — quam illi ponant, qui ponunt eum tantum partem animae. Est
tamen in anima rationali aliquid activum, quod dico potentiam illam, per quam nata est se in
omnium intelligibilium similitudinem transformare. Haec igitur vis, si appelletur intellectus
agens — quare et habet aliquid luminis creati naturalis — differt essentialiter ab intellectu
possibili, sicut vis a vi, non sicut potentia a potentia. Credo enim quod sunt diversae vires
eiusdem potentiae, sicut in eodem organo oculi differunt splendor et perspicuitas. Ratio
autem inferior et superior dicunt eandem diversimode relatam. Dico ergo quod intellectus
agens creatus suo modo perficit utramque. ... intellectus agens non copulatur cum
superioribus vel inferioribus nisi per possibilem. ... Intellectus autem possibilis natus est
utrobique, et a superiori et ab inferiori, illustrari." *Ibid.*, Q. 7, Resp., *ed. cit.*, pp. 73-4.

> their parents. ... The foundation of this error [that there is
> one intellect for all men] must be identified so that the error
> may be destroyed at the foundation by the testimonies of
> the saints founded in eternity.[34]

He then launches an attack on the "seminator huius erroris" under three heads, the substance of the possible intellect, its powers, and its operations, using the principles we have just noted to refute Averroes. He uses the doctrine of the matter-form composition of the soul to confute those who maintain the uniqueness of the intellect. The substance of the possible intellect is not free of all matter, for if it were it could not be any particular being and consequently could not be the perfection of man in the same way that the sensitive soul is, but equivocally. The powers of the possible intellect are both active and passive. Concerning its operation, Averroes contradicts Aristotle, who considers understanding to occur in the individual man. According to Averroes, "reason does not place man in his species, nor is it his perfection in first being. From this it follows that man is not to be defined through rational."[35] This position "destroys the foundations of nature, as well as merits and rewards. ... Therefore this heresy is to be repudiated. No one assents to it but a pernicious heretic, and no one defends it as probable but a madman, completely ignorant of both divine and human literature."[36]

The battle was clearly heating up. But Pecham, despite his zeal, was not a worthy opponent on the philosophical level for the young artists. He holds two positions that seem to be mutually incompatible: first, that the soul is the substantial form of the body; and second, that the soul is composed of matter and form. He adopted the first position in order to preserve the unity of the individual human being and his dignity as a rational substance. He adopted the second in order to explain how the soul could be a *hoc aliquid* and be multiplied according to the number of human beings. In order to save its incorruptibility, even though it contained matter, he adopted Avicebron's

---

[34] "Sicut tradit fides catholica et sanctorum sententiae contestantur, una est cuius-libet hominis singularis anima rationalis, quare corpori creato infunditur mox creata. Nec umquam sanctus aliquis de sententia hac dubitavit, quamvis de origine animarum constet aliquos dubitasse, utrum scilicet singularum animae creentur de nihilo an a parentibus traducantur. ... Videnda igitur sunt fundamenta huius erroris, ut error destruatur in fundamentis sanctorum testimoniis in aeternum fundatis." *Ibid.*, Q. 4, Resp., *ed. cit.*, p. 49.

[35] "... non ponit ratio hominis in specie, nec est eius perfectio in esse primo. Ex quo sequitur quod non esset definiendus homo per rationale." *Ibid.*, Q. 4, Resp., *ed. cit.*, p. 52.

[36] "Haec igitur positio destruit fundamenta naturae, item merita et praemia. ... Haec igitur haeresis est repudianda. Cui nullus assentit nisi perniciosus haereticus; defendit ut probabilem nisi phreneticus et litterarum divinarum et humanarum penitus ignarus." *Loc. cit.*

doctrine of spiritual matter and form, although he attributed it to Augustine and cited several Augustinian texts,[37] which, when read in the light of Avicebron, could be interpreted in this way. Pecham seems to have been indebted to William of Auvergne for his doctrine of the agent intellect; on one occasion he identifies it with God, the uncreated *lumen* of which Augustine speaks, but he insists that the rational soul has both an active and a passive aspect, although he does not pursue the problems inherent in this position with the same perseverence William had shown.

With William of Baglione and John Pecham we have reached the point where the dangers inherent in Averroes have been fully realized and the counterattack has been launched. But the tradition these anti-Averroists appealed to is of their own very recent creation. The doctrines of the Fathers, especially Augustine, have been radically altered by having been re-interpreted in the light of Muslim, Jewish and Greek authors.

SIGER OF BRABANT, *IN TERTIUM DE ANIMA*. Up to this point, we have been dependent upon their adversaries for information about what the 'Averroists' were teaching. But we are fortunate in possessing, in an excellent critical edition, Siger of Brabant's *In tertium De anima*,[38] a work universally conceded to have been composed in or before 1269. It is generally characterized as being a purely Averroistic work, exemplifying Siger's position before Aquinas's *De unitate intellectus* forced certain modifications upon him. While it is true that in this work Siger maintains the uniqueness of the intellect, his stance is not that of a belligerent rationalist ridiculing the superstitions of the theologians; even at this early stage of his career he exhibits an accommodating disposition.

He begins his work by specifying his doctrine of what the soul is (which is not strictly Averroistic). The vegetative and sensitive soul, he says, is educed from the potency of matter. But the intellective soul, which comes 'from without,' is not rooted in the same simple soul as the vegetative and sensitive, nor does it possess any vegetative or sensitive functions. If it did, there would have to be a double vegetative and sensitive soul in humans. When the intellective soul does arrive from without, it does not corrupt the already-existing vegetative and sensitive soul—since a thing is corrupted only by its

---

[37] *De Genesi contra Manichaeos* I, 6; *Confessiones* XII, 8; and *De Genesi ad litteram* V, 5.

[38] Bernardo Bazán, ed., *Siger de Brabant. Quaestiones in Tertium De Anima, De Anima Intellectiva, De Aeternitate Mundi.* Philosophes Médiévaux 13 (Louvain/Paris, 1972).

contrary—but is united to them. "And this united soul does not constitute one simple soul, but one composite soul."[39]

In his discussion of whether the intellect is eternal or created *de novo*, Siger displays exemplary moderation. First he presents Aristotle's position: according to Aristotle, one must say that the intellect is eternal, since everything that is made immediately by the First Cause is not new but is an eternally made thing. So Aristotle's judgment is that the intellect, like the world, is eternal. And the intellect, which is the mover of the human species, is one eternal made thing and is not multiplied by the multiplication of individuals. The reason Aristotle says this, he explains, is that every agent that makes something *de novo* is altered in the process. Therefore, if the First Cause made something *de novo*, there must have been some novelty and alteration in his will.

Siger then asks if Aristotle's position is necessary, and he answers that it is not necessary, but only probable. In order to have a necessary demonstration, we should need to know the form of the divine will. "But who would be able to do that?"[40] The only answer to why the intellect was made eternal or *de novo* is that God so willed it. And his will does not depend on external events, such as our will does, nor can it be coerced or even influenced by anything. And Aristotle's reason for holding that everything made *de novo* causes a change in the agent does not apply in this case. It would be true only of an agent that did not act by the form of its will. "But because it is not true here—because the intellect was made by an agent according to the form of its will—then it is not necessary that its being made *de novo* would require a change in the agent."[41] Aristotle's position, however, is more probable than that of Augustine, and one may not accept both positions at the same time; if one is right, the other is wrong.

In question 4, "Whether the intellect is generable," Siger proposes a solution to the psychological problem apparently posed by monopsychism, which had been advanced by several previous authors and which Aquinas would develop further in his *De unitate intellectus,* and in his answer Siger shows his

---

[39] "... et sic ipsa unita non faciunt unam simplicem, sed compositam." *In tertium De anima,* Q. 1, solutio, *ed. cit.,* p. 3. Three lines later he says: "Verum est: unam compositam, non autem unam simplicem."

[40] This same view figures largely in Boethius of Dacia's *De aeternitate mundi.* See Richard C. Dales, "Maimonides and Boethius of Dacia on the Eternity of the World," *The New Scholasticism* 56 (1982), 306-19, on p. 313.

[41] "Sed quia non est ita hic, quia intellectus est factus ab agente quod agit suae voluntatis, tunc non oportet quod sit factum ab agente transmutato." Q. 2, solutio, *ed. cit.,* p. 8.

fundamental differences both with Albert and Thomas and with Boethius of
Dacia. First he presents the Averroistic account of intellection: only the
intellect is free of matter and can apprehend universal forms. A form that is
joined to matter in its substance and is educed from the potency of matter uses
an organ or instrument. But the action and operation of the intellect is
separate and does not use an organ. Whatever the imagination or sense
apprehend, they grasp under material conditions. We understand by the
abstraction of intelligible forms from the phantasms, but in the very act of
understanding the intellect does not use an organ.

But how do we experience this act of understanding to be brought about in
us by the intellect? Isn't this operation the sole property of the intellect?
Siger answers that in a certain way it is.

> For we are conscious of the intellect through the powers of
> the body. And we perceive those operations that are in us,
> or are brought about in us, by means of the body and
> matter; and similarly we perceive the operations that are
> brought about in us by reason of the intellect. Whence it is
> itself our intellect, through which we experience this kind
> of universal reception to be accomplished in us. For our
> intellect grasps itself as an operation.[42]

But in order for the intellect to behave in this way, it is necessary that it have
in it some sort of potency, or something that could act in a manner analogous
to matter, since it must be in potency to all intelligible forms. In question 6
Siger asks whether the soul is composed of matter and form. He points out
that "in the intellect there is some potency, since it is not pure act at the limit
of simplicity, in which there is not any composition. ... But all other things
[than God], which recede from his simplicity, receive something of
composition. ... Therefore, since the intellect recedes from the pure act and
simplicity of the First, it must have some sort of composition."[43] But this is
not composition of matter and form, and those who hold the doctrine of
universal hylomorphism are mistaken. Rather, it is a compound form, such as

---

[42] "Nos enim conscii sumus ex virtutibus corporis intellectum. Et percipimus
operationes quae in nobis sunt vel fiunt ratione virtutum corporis et materiae, et similiter nos
percipimus operationes quae fiunt in nobis ratione intellectus. Unde ipse est intellectus
noster, per quem experimur huiusmodi acceptionem universalem fieri in nobis. Intellectus
enim noster apprehendit se ipsum sicut operari." Q. 4, solutio, *ed. cit.*, p. 14.

[43] "... in intellectu sit aliqua potentia, cum non sit actus purus in fine simplicitatis,
in quo non est aliqua compositio. ... Alia vero omnia, quae a sua simplicitate recedunt,
compositionem aliquam recipiunt. ... Ideo cum intellectus a puro actu Primi recedat et
simplicitate, oportet quod aliquam compositionem habeat." Q. 6, solutio, *ed. cit.*, pp. 18-19.

a genus and the form of the differentia. "For not all forms are simple. Since all the parts of a definition are forms, it must be that one of these be material with respect to the other."[44]   But even though the intellect is form without matter, it is not the substantial form of the body. It perfects the body not through its substance, but through its power, because if it perfected it by its substance, it would not be separable. And if it were the substantial form of the body, there would be no question as to whether the intellect were one or many. It would clearly be many. But because it lacks matter and is separated, it is one. It is joined to us because it is in potency to the senses, imaginations, and phantasms that are produced by the body. The contention that if the intellect were one, when one man acquired knowledge, all men would, is vain, because there is no understanding without the data provided by the senses and imagination. It is these upon which the agent intellect acts. "Therefore, since it is not necessary that if one person imagines something, so does another, and if one doesn't neither does the other, consequently if one person acquires knowledge, another need not also acquire it."[45]   Nor is there any innate knowledge even of certain first principles, such as Albert the Great contends, which, although they are not the agent intellect, act as its instruments. "I say and believe that there is no cognition of intelligibles innate to our intellect, but it is in pure potency to all intelligibles, having interior innate knowledge of none of them; but from the phantasms it understands whatever it understands."[46]   In this same question, Siger repeats Philip the Chancellor's characterization of man as the most composed of all creatures, a point that had also been adopted by John of La Rochelle.

We may summarize the doctrine of this treatise as follows: The intellective soul, insofar as it is our intellect, is a compound soul, a union of the vegetative-sensitive soul with the one intellect. This union occurs because the intellect is by its nature in potency to all intelligible forms, which are supplied by the body. It is not united to the body through its substance, but only through its operation, and only in this sense is it the body's perfection. It completely lacks matter but does have some composition, which enables it to be receptive of the data provided by the body. The one intellect is probably,

---

[44] "Non enim omnes formae simplices sunt. Cum enim partes omnes definitionis formae sint, oportet quod unum sit materiale respectu alterius." Q. 6, solutio, *ed. cit.*, p. 21.

[45] "Quare, cum non sit necesse quod, si unus imaginetur, quod alius, quod si unus <non>, quod alius, nec per consequens, si unus acquirat scientiam, quod alius." Q. 9, solutio, *ed. cit.*, p. 29.

[46] "Dico et credo quod intellectui nostro non est innata aliqua cognitio intelligibilium, sed est in pura potentia ad omnia intelligibilia, nullius intus habens innatam cogitionem, sed ex phantasmatibus intelligit quidquid intelligit." Q. 12, solutio, *ed. cit.*, p. 40.

but not necessarily, eternal rather than *de novo*, not by its own nature (which requires neither the one nor the other), but solely because of God's will. The intellect is united to us by its very nature, and we experience understanding as an operation worked upon the sensibles provided by the body. The agent of this operation is separate, but the fact of understanding itself is the result of the cooperation of the phantasms and the intellect and thus shares in matter and the body.

This is a carefully thought out treatise in the form of questions, probably reflecting a course in the Parisian arts faculty. It can hardly be characterized as teaching boys in corners.

It will be noticed that in this treatise, quite apart from the Averroistic interpretation of the possible intellect, there is much that is familiar, and that Siger's thought is firmly in the tradition of earlier scholastic thought on the soul. As a professor of philosophy, he is necessarily obliged to use the term 'form' in its Aristotelian sense. He denies that the soul is the substantial form of the body because, in order to be so, it would have to be joined to matter in order to exist. In emphasizing its operational, rather than substantial, relation to the body, he is very close to Albert the Great. In denying that the vegetative and sensitive souls are of the same substance as the intellective, he agrees with the majority of the thinkers we have examined, although he disagrees with Albert. In holding that the human soul is a composite one, he is firmly in the tradition of the Parisian artists from Philip the Chancellor (himself a theologian) onward; and from Philip too he borrowed the doctrine that man is the most composed of all creatures. In understanding the composition of the soul to be of forms, one of which behaves as matter to the other, he is reasserting a point that had previously been made by John Blund, Peter of Spain, and Anonymous Vaticanus. And in denying that the rational soul would corrupt, rather than complete, the sensitive soul, he had predecessors in Anonymous Admontensis and Anonymous Vaticanus. What was truly revolutionary in Siger's teaching was his adoption of Averroes's interpretation of Aristotle's doctrine of the possible intellect to be the correct (but not necessarily true) one: the intellect must be free of matter in order to perform its proper function, and if it is free of matter it cannot be individuated; and therefore there can be only one agent and possible intellect.

Whether this means that there is one soul for the human race is a more complicated question. Siger would later modify his doctrine of the soul considerably, but is seems to me that even in this early work, Siger considers each man to have his own soul, compounded of the individual sensitive soul and the unique rational soul. Although there are grave difficulties with this position, they are not more serious than with any position that posits a compound soul. Both, from the standpoint of Christianity, would have to be considered heretical, though for different reasons, if taught as the truth. But

Siger's purpose in this commentary was confined to interpreting correctly the teaching of Aristotle.

CHAPTER SEVEN

AQUINAS'S *DE UNITATE INTELLECTUS*

By 1269 we may discern three general groupings of doctrines on the soul. One was derived from the teaching of Averroes, and its only novelty was the assertion that Aristotle had taught that the possible intellect is one for all humans. Another was that of Aquinas, who had denied the body-soul duality and taught that these two were intimately united as matter and form, constituting one substance; the soul was individuated by virtue of having been created to be the form of an individual body and was able to retain its individuality and being after the dissolution of the body, but it was only complete when it was acting as the body's form. A third doctrinal complex had recently been devised by William of Baglione and John Pecham, building somewhat differently on the thought of Bonaventure, primarily to combat the threat posed by monopsychism but also to avoid what they considered to be the consequences (almost as bad) of Aquinas's teaching, especially his views on the unity of substantial form, the simplicity and non-materiality of the soul, the soul's individuation by the body, and its not being strictly speaking a *hoc aliquid*. To strengthen the soul's substantiality and to guarantee its immortality, they adopted the universal hylomorphism of Avicebron and reinterpreted Augustine's views in light of it. But this necessitated their abandoning a genuine Augustinian position—the unity and simplicity of the soul—to adopt instead a theory of plurality of forms, which had earlier been developed by the Aristotelians. Since Averroes denied that the intellect (and the Latins most often understood this as 'rational soul') was the substantial form of man, they insisted upon the formula, a circumstance that required them to define 'form' in a non-Aristotelian way; they usually followed Avicenna in their definition.

In this year too, whether to deal with the serious disagreements in the arts and theological faculties or, more likely, to deal with the renewed outbreak of hostilities between seculars and mendicants, the Dominican order sent Thomas Aquinas from Naples, where he had been working on commentaries on Aristotle's *Physics* and *Metaphysics*, to Paris for his second regency in theology. One of the first fruits of Thomas's return to Paris was his treatise *De unitate intellectus*.

After his introduction explaining the purpose of the treatise, Thomas first treats the question of whether the intellect is a substance separate in its being from the body. He argues first that this is not Aristotle's view, second that it is not the view of the Greek and Arabic commentators, and third that it is not consistent with sound philosophy or with our experience of understanding. Next he investigates the contention that the intellect is one for all men, a proposition that he attempts to prove is not Aristotle's view and is not true. The work ends with a rebuke to the master or masters who teach these doctrines and a challenge to them to commit their teaching to writing.

Both at the beginning and the end of this work, Thomas presents us with some hints of the context of its composition. He begins by saying that a particularly serious error, stemming from Averroes, has been spreading among many people for some time—namely that the possible intellect is: 1- "a certain substance separate from the body in its being and not united to it in some way as its form;"[1] and 2- that the possible intellect is one for all men. He goes on to say that since the boldness of those who maintain this position has not ceased, despite Thomas's earlier writings on the subject, he will try to refute the error more clearly. And at the end of the work he uses language that can be (but need not be) understood as referring to a specific individual, who had been teaching in corners to boys who are too young to evaluate his doctrine. Thomas implies that his knowledge of this teaching does not come from a written work when he challenges his adversary (or adversaries) "to write against this treatise if he dares." This work was not then, as was Albert the Great's *De unitate intellectus*, written only against Averroes; it was also written against those Parisian professors who were teaching the Averroistic interpretation of Aristotle's *De anima*. And although he directs most of the treatise against monopsychism, Thomas also takes shots at other Parisian theologians who had opposed some of his views.

Thomas says that he will conduct his inquiry on a purely philosophical level (although he does not manage to keep this promise). He is concerned to show what Aristotle truly taught on the two points in question, to show how Averroes misinterpreted him (and hence can be called a perverter of Aristotle), and to show what the truth of the matter is.

He first attempts, in chapters 1 through 3, to show that Aristotle did not teach that the possible intellect is a substance separate in its being from the body. He accepts Aristotle's definition of the soul from *De anima* 2, 1 (412b): it is the first act, i.e., the substantial form, of a physical organic body. The intellect is included in this definition, Thomas says, and he cites various other texts of Aristotle to prove that this is what he intended, and then to show that Averroes was wrong in understanding him to have taught that the intellect is not the soul. When Aristotle said that "it belongs to the intellect alone to be separated, as the eternal from the corruptible," he did not mean, as Averroes

---

[1] "... quandam substantiam secundum esse a corpore separatam, nec aliquo modo uniri ei ut forma." Aquinas, *De unitate intellectus*, in Sancti Thomae de Aquino *Opera omnia* iussu Leonis XIII P. M. edita, XLIII (Rome , 1976), p. 291a.

(and indeed most Latin scholastics) understood him, that the vegetative and sensitive powers were inseparable from the body, but that the various powers of the soul are not separable from each other, but are distinct only by reason; and the soul as a whole is united to the body not as a pilot to a ship, but as a form. And Thomas clinches his argument by quoting Aristotle's words in *De anima* 3, 4 (429a): "For he says: 'Concerning the part of the soul by which the soul thinks and understands.'"

Thomas then marshals a number of arguments to show that Aristotle taught that "the intellect is a power of the soul, which [soul] is the act of the body." In this regard, he considered a question that was not confined to the Averroists, but was also maintained by the right wing of the modernists, William of Baglione and John Pecham: "How is it possible that the soul is the form of the body, and some power of the soul is not a power of the body?"[2] Thomas argues that we see instances of a form that is the act of a body made of the elements, which has a power that does not belong to any of the elements but which it possesses by virtue of a higher principle, such as a celestial body; the more noble the form, the more it has powers that transcend matter. And the highest of all forms, the human soul, has a power, namely understanding, that totally transcends matter. The intellect then is separate because it is not a power of the body, but is a power of the soul. "We do not say that the soul, in which the intellect is, so exceeds matter that it does not have its being in the body, but that the intellect, which Aristotle calls a power of the soul, is not the act of the body."[3] It is the soul itself that is the act of the body and gives it specific being. Some of its powers are the act of certain parts of the body, but the intellect is a power that is not, because its operation is not accomplished by means of a bodily organ. The form of man therefore is both in matter and is separate. It is in matter according to the being it gives to the body, but separate according to the power that is proper to man, i.e., the intellect. "It is therefore not impossible that some form be in matter, and its power be separate."[4] This point is central to Thomas's doctrine and is the point on which both Siger and William of Baglione attacked him. But his presentation of the position here is not especially strong.

---

[2] "Quomodo autem hoc esse possit, quod anima sit forma corporis et aliqua virtus anime non sit corporis virtus." *Ed. cit.*, p. 296b.

[3] "Nec dicimus quod anima, in qua est intellectus, sic excedat materiam corporalem quod non habeat esse in corpore; sed quod intellectus, quem Aristotiles dicit potentiam anime, non est actus corporis. *Ed. cit.*, p. 296b.

[4] "Non est ergo impossibile quod aliqua forma sit in materia, et virtus eius sit separata." *Ed. cit.*, p. 297a.

Thomas shifts his focus here from his own philosophical presentation to an analysis of what Aristotle says on the same point. There is no doubt that Aristotle taught that that the intellect was incorruptible. How then can it be the form of a corruptible body? Thomas cites *Metaphysics* 12, 3 (1070a) to show that whereas Aristotle said forms do not exist prior to matter, nothing he says prevents a form such as the intellective part of the soul from remaining after matter. Aristotle says of the intellect that "only when it is separated is it what it truly is, and this alone is immortal and eternal." Thomas admits that there is some doubt as to what Aristotle is talking about here but shows to his own satisfaction that it is "the whole intellective part, which indeed is called separate because it has no organ."

Thomas then distinguishes between forms that exist only in being joined to matter and have no operation except through the composite thus formed, and forms that have some power of their own apart from being joined to matter. The latter do not exist only through the being of the composite, but rather the composite exists through the form's being, and so when the composite is destroyed, it is not necessary that such a form be destroyed along with it. In discussing how the soul will have an intellectual operation after it is separated from the body, Thomas resorts to the same evasion that Siger did—that the question does not apply to natural philosophy. But Aristotle implies that its way of knowing will be different then than it is now when he asks, *De anima* 3, 7 (431b) whether the intellect that is not separate from the body may know something separate.

Therefore it is clear that the arguments to the contrary are not necessary demonstrations, says Thomas. It is essential to the soul that it be united to a body. But this is hindered accidentally, not because of the soul, but because of the body that is corrupted, just as a light thing may be accidentally hindered from being up.

The final point Thomas treats in chapter 1 is the very difficult question of how the vegetative and sensitive powers, corruptible and educed from the potency of matter, and the intellectual soul, incorruptible and from without, can be one soul. Exploiting Aristotle's analogy of geometrical figures, as several of his predecessors had done, Thomas gives a clearer though similar solution, although it holds only if one accepts Thomas's account of the progression of forms: a square is other in species than a triangle, but not other than the triangle that is potentially in it, since the two have the same producing cause.

> Thus therefore the vegetative soul existing apart from the sensitive is indeed another species of soul and has another producing cause; nevertheless there is the same producing cause for the sensitive and the vegetative that is within the sensitive. If therefore it be stated thus: that the vegetative

and sensitive [power] that is present within the intellective
is from the extrinsic cause that is the cause of the
intellective, nothing illogical results. For it is not illogical
for the effect of a superior agent to have a power that the
effect of an inferior agent has, plus something more.
Whence also the intellective soul, although it be from an
external agent, nevertheless has powers that the vegetative
and sensitive souls, that are from inferior agents, also
possess.[5]

And the chapter ends with the re-assertion that "it is clear that [Aristotle's]
position was that the human soul is the act of the body, and that the possible
intellect is a part or power of that soul."

The next chapter takes Averroes, and his followers in this matter, to task for
claiming that all the Peripatetics had understood Aristotle as having taught
that the intellect was a substance separate from the body. Thomas had
recently been able to read Themistius's *Paraphrasis eorum quae de anima
Aristotelis* in the Latin translation that his Dominican confrère, William of
Moerbeke, had made, and Thomas noticed that Averroes had misrepresented
Themistius's teaching on this point, a circumstance of which he made much.
The import of this chapter is that many commentators on *De anima*, both
Greek and Arabic, had interpreted Aristotle as teaching that the intellect was
part of the soul of individual human beings.

Chapter 3 considers the same question from the standpoint of philosophy.
Thomas again begins with the text of *De anima* and states the heart of his
argument thus:

For it is clear that this individual man understands, for we
would never inquire about the intellect unless we
understood; nor when we inquire about the intellect, do we
ask about any other principle than that by which we
understand. Wherefore also Aristotle says: "Moreover, I
mean the intellect by which the soul understands." And
Aristotle concludes thus, that if something is the first
principle by which we understand, that [principle] must be

---

[5] "Sic igitur uegetatiuum quidem seorsum a sensitiuo existens alia species anime est,
et aliam causam productiuam habet; eadem tamen causa productiua est sensitui, et vegetatiui
quod inest sensituo. Si ergo sic dicatur quod vegetatiuum et sensitiuum quod inest
intellectiuo, est a causa extrinseca a qua est intellectiuum, nullum inconueniens sequitur: non
enim inconueniens est effectum superioris agentis habere uirtutem quam habet effectus
inferioris agentis, et adhuc amplius; unde et anima intellectiua, quamuis sit ab exteriori
agente, habet tamen uirtutes quas habent anima uegetatiua et sensitiua, que sunt ab
inferioribus agentibus." *Ed. cit.*, pp. 300b-301a.

the form of the body, because he has previously made it clear that that by which anything first of all operates is the form. And this is evident through the reason that each thing acts insofar as it is in act; but each thing is in act through form. Therefore it must be that that by which something first acts is the form.[6]

The crux of his argument is that according to Averroes's explanation of intellection, it would not be possible to say that this individual man understands, a point that had earlier been made by William of Baglione. He first criticizes the text of Averroes himself, arguing that the contact between the phantasms and the separate intellect would not be sufficient for us to say that a man understands. Then he investigates just what it is that constitutes a man (Socrates in this case), and ends with the summary:

> But if it be said that this individual, who is Socrates, is a body animated by a vegetative and sensitive soul, as seems to follow according to those who posit that this man is not constituted in his species through the intellect but through the sensitive soul ennobled by some illumination from, or coupling with, the possible intellect, then intellect behaves toward Socrates only as a mover to a thing moved. But according to this, the action of the intellect, which is understanding, could in no way be attributed to Socrates.[7]

Then assuming it as given that 'this man understands,' Thomas shows that if one grants that the intellect is a separate substance and not related to man as a form, there is no way it is true that this man understands, since understanding is an action of the intellect only. It would also follow, from the assumption

---

6 "Manifestum est enim quod hic homo singularis intelligit: numquam enim de intellectu quereremus nisi intelligeremus; nec cum querimus de intellectu, de alio principio querimus quam de eo quo nos intelligimus. Unde et Aristotiles dicit "Dico autem intellectum quo intelligit anima." Concludit autem sic Aristotiles quod si aliquid est primum principium quo intelligimus, oportet illud esse formam corporis; quia ipse prius manifestauit quod illud quo primo aliquid operatur est forma. Et patet per rationem, quia unumquodque agit in quantum est actu; est autem unumquodque actu per formam: unde oportet illud quo primo aliquid agit esse formam." *Ed. cit.* p. 303a.

7 "Si vero dicatur quod hoc indiuiduum quod est Sortes, est corpus animatum anima uegetatiua et sensitiua, ut uidetur sequi secundum eos qui ponunt quod hic homo non constituitur in specie per intellectum, sed per animam sensitiuam nobilitatam ex aliqua illustratione seu copulatione intellectus possibilis: tunc intellectus non se habet ad Sortem nisi sicut mouens ad motum. Sed secundum hoc actio intellectus que est intelligere, nullo modo poterit attribui Sorti." *Ed. cit.*, p. 304b.

that the intellect is separate from the body, that the will, which is in the intellect, would also be separate. In this case, a man would not be responsible for his actions, and so the principle of moral philosophy would be destroyed. Therefore, the intellect must be "united in such a way that from it and from us is made truly one thing, which surely can only be in the way in which it has been explained, that is, that [the intellect] is a power of the soul that is united to the body as form. It remains therefore that this must be held without any doubt, not on account of the revelation of faith, as they say, but because to deny this is to strive against what is clearly apparent."[8]

But Thomas was aware of the criticism of the position that the intellect would thereby be a material form and as a result not be able to strip sensible phantasms of their material and particular circumstances. He reiterates his argument from *Summa contra Gentiles* 2, 60:

> For we do not say that the human soul is the form of the body according to its intellective power, which according to Aristotle's teaching is not the act of any organ. Whence it remains that the soul, as regards its intellective power, be immaterial, both receiving things immaterially and understanding itself. Whence also Aristotle expressly says that the soul is the place of species, "not the whole [soul], but the intellect." But if it is objected against this that a power of the soul cannot be more immaterial or more simple than its [the soul's] essence, the argument would proceed soundly if the essence of the human soul were the form of matter in such a way that it would not exist through its own act of existing, but through the act of existing of the composite, as is true of other forms, which of themselves have neither the act of existing nor any operation apart from their union with matter; which for this reason are said to be immersed in matter. But the human soul, because it exists by its own act of existing, in which matter shares to some extent [though] not wholly comprising it, since the dignity of this form is greater than the capacity of matter. Therefore nothing prevents the soul from having some operation or power to which matter cannot attain.[9]

---

[8] "... intellectus sic uniatur nobis ut uere ex eo et nobis fiat unum; quod uere non potest esse nisi eo modo quo dictum est, ut sit scilicet potentia anime que unitur nobis ut forma. Relinquitur igitur hoc absque omni dubitatione tenendum, non propter reuelationem fidei, ut dicunt, sed quia hoc subtrahere est niti contra manifeste apparentia." *Ed. cit.*, p. 306b.

[9] "Non enim dicimus animam humanam esse formam corporis secundum intellec-tiuam potentiam, que secundum doctrinam Aristotilis nullius organi actus est: unde remanet

Chapter 4 takes up the second question, whether the possible intellect is one for all men. Thomas concedes that no difficulty seems to follow from the assumption that the agent intellect is one for all men, since many things can be perfected by one agent; but, he says, this is not the meaning of Aristotle. However, it is impossible that there be one possible intellect for all men. First, because, if the possible intellect is that by which we understand, then it is necessary to say that the individual man who understands either is the intellect itself, or that the intellect, which is a power of the soul that is the form of the body, inhere in him formally. In the first case, this individual man would not be different from that individual man, and all men would be one individual man. In the second case, any line of reasoning one might pursue leads to the conclusion that the intellect is not unique but is distinct for each individual, and this in fact is Aristotle's position.

This is followed by an ingenious argument, which both Siger and Anonymous Van Steenberghen found compelling. We must note three things as established, says Thomas. First, the *habitus* of knowledge is the first act of the possible intellect itself, which according to this *habitus* comes into act and can act through itself. But knowledge is not only according to the illuminated phantasms, as some say, nor is it a faculty that is acquired by us from frequent meditation or exercises, so that we may be in contact with the possible intellect through our phantasms. Second, before our learning or discovering, the possible intellect itself is in potency like a tablet on which nothing is written. And third, by our learning or discovering, the possible intellect is put into act. But these three propositions are inconsistent with a single intellect for all men who have been, are, and will be.

> For it is clear that the species are retained in the intellect (for it is the place of species, as the Philosopher had said above). And again, knowledge is a permanent *habitus*. If therefore through some previous man, the intellect has been put into act according to some intelligible species and perfected according to the *habitus* of knowledge, that

---

quod anima, quantum ad intellectiuam potentiam, sit immaterialis et immaterialiter recipiens et se ipsam intelligens. Vnde et Aristotiles signanter dicit quod anima est locus specierum "non tota sed intellectus." Si autem contra hoc obiciatur quod potentia anime non potest esse immaterialior aut simplicior quam eius essentia: optime quidem procederet ratio, si essentia humane anime sic esset forma materie, quod non per esse suum esset sed per esse compositi, sicut est de aliis formis, que secundum se nec operationem habent preter communicationem materie, que propter hoc materie immerse dicuntur. Anima autem humana, quia secundum suum esse est, cui aliqualiter communicat materia non totaliter comprehendens ipsam, eo quod maior est dignitas huius forme quam capacitas materie: nichil prohibet quin habeat aliquam operationem vel uirtutem ad quam materia non attingit." *Ed. cit.*, p. 307a.

habitus and those species remain in it. But since every recipient must be lacking that which it receives, it will be impossible that through my learning or discovering those species be acquired in the possible intellect. For even if someone should say that through my discovery the possible intellect would be put into act regarding something new—for example, if I discover some intelligible that has been discovered by no previous man—nevertheless this cannot happen in learning, for I can only learn what one who teaches me has known. Therefore it is vain to say that before learning or discovering the intellect was in potency.

But if someone should add that men always existed, according to the opinion of Aristotle, and that therefore there would not have been a first man understanding, and so through no one's phantasms have the intelligible species been acquired in the possible intellect, but the intelligible species of the possible intellect are eternal. In vain therefore did Aristotle posit the agent intellect, which would make intelligibles in potency to be intelligibles in act. In vain too did he posit that phantasms behave toward the possible intellect as colors to sight if the possible intellect gets nothing from the phantasms. Besides, it would seem to be very unreasonable that a separate substance should receive [something] from our phantasms and that it would not be able to know itself except after our learning or understanding, because Aristotle adds after the foregoing words: "And it can then understand itself," that is, after learning or discovering. For a separate substance is intelligible in itself; therefore, the possible intellect, if it were a separate substance, would understand itself through its own essence, nor it would it need for this intelligible species that would come to it through our understanding or discovery.[10]

---

[10] "Manifestum est enim quod species conseruantur in intellectu, est enim locus specierum, ut supra Philosophus dixerat; et iterum scientia est habitus permanens. Si ergo per aliquem precedentium hominum factus est in actu secundum aliquas species intelligibiles, et perfectus secundum habitum scientie, ille habitus et ille species in eo remanent. Cum autem omne recipiens sit denudatum ab eo quod recipit, impossibile erit quod per meum addiscere aut inuenire ille species acquirantur in intellectu possibili. Etsi enim aliquis dicat quod per meum inuenire intellectus possibilis secundum aliquid fiat in actu de nouo, ita si ego aliquid intelligibilium inuenio quod a nullo precedentium est inuentum: tamen in addiscendo hoc contingere non potest, non enim possum addiscere nisi quod docens sciuit. Frustra ergo dixit quod ante addiscere aut inuenire intellectus erat in potentia. Sed et si quid addat homines semper fuisse secundum opinionem Aristotilis, sequetur quod non fuerit primus homo intelligens; et sic per fantasmata nullius species intelligibiles sunt acquisite in intellectu possibili, sed sunt species intelligibiles intellectus possibilis eterne. Frustra ergo Aristotiles

And after arguing that Aristotle was not talking about the possible intellect only insofar as it was in contact with us and not as it is in itself, Thomas concludes this section: "So therefore in every way it is impossible that there should be only one possible intellect for all men."

Having made his own case, Thomas turns in chapter 5 to a refutation of the arguments against the plurality of the possible intellect. The first of these is that the intellect cannot be many because it is free of matter. In answering this, he repeats from his *Commentary on the Sentences* 4, d. 12, q. 1, ad 1 et 3 and *Summa theologiae* 3, q. 77, art. 2, showing in what ways matter is and is not the principle of individuation:

> For matter is not the principle of individuation in material things except insofar as matter cannot be participated in by several things, since it is the first subject not existing in another. Whence also concerning the [Platonic] idea, Aristotle says that if the idea were separate, "it would be something, that is, an individual, which could not possibly be predicated of many." Separate substances therefore are individuals and singular. But they are not individuated by matter, but by the very fact that it is not their nature to be in another, and consequently are not to be participated in by many either. From this it follows that if it is the nature of some form to be participated in by another in such a way that it is the act of some matter, that [form] can be individuated and multiplied through relation to matter. But it has already been shown above that the intellect is a power of the soul, which is the act of the body. In many bodies therefore there are many souls, and in many souls there are many intellectual powers that are called intellects. Nor for this reason does it follow that the intellect would be a material power, as was shown above.[11]

---

posuit intellectum agentem, qui faceret intelligibilia in potentia intelligibilia in actu; frustra etiam posuit quod fantasmata se habent ad intellectum possibilem sicut colores ad uisum, si intellectus possibilis nichil a fantasmatibus accipit. Quamuis et hoc ipsum irrationabile uideatur, quod substantia separata a fantasmatibus nostris accipiat, et quod non possit se intelligere nisi post nostrum addiscere aut intelligere; quia Aristotiles post uerba premissa subiungit "et ipse se ipsum tunc potest intelligere," scilicet post addiscere aut inuenire. Substantia enim separata secundum se ipsam est intelligibilis: unde per suam essentiam se intelligeret intellectus possibilis, si esset substantia separata; nec indigeret ad hoc speciebus intelligibilibus ei superuenientibus per nostrum intelligere aut inuenire." *Ed. cit.*, p. 309a-b.

[11] "Non enim materia est principium indiuiduationis in rebus materialibus, nisi in quantum materia non est participabilis a pluribus, cum sit primum subiectum non existens in alio; unde et de ydea Aristotiles dicit quod, si ydea esset separata "esset quedam, id est indiuidua, quam impossibile esset predicari de multis." Indiuidue ergo sunt substantie

And in arguing against the contention that the multiplicity of the intellect would involve a contradiction, Thomas enunciates the principle upon which Boethius of Dacia's argument in *De aeternitate mundi* was based: "So therefore if the intellect were naturally one for all men because it would not have a natural cause of multiplication, it could nevertheless receive multiplication from a supernatural cause, and this would not imply a contradiction."[12]

In replying to the objection that if a plurality of intellectual substances remained after the destruction of their bodies, they would be idle, Thomas restates his position from *De spiritualibus creaturis* 2, ad 5:

> But we concede that the human soul, when separated from the body, does not have the highest perfection of its nature, since it is part of human nature. For no part has complete perfection if it is separated from the whole. But it does not, for this reason, exist in vain, for the end of the human soul is not to move a body, but to understand, in which is its happiness, as Aristotle proves in book 10 of the *Ethics*.[13]

Then once again Thomas refutes the contention that the uniqueness of the intellect was asserted by all the philosophers except the Latins, dwelling especially on Averroes's misrepresentation of the doctrine of Themistius, and concludes: "Hence we have for good reason called him [Averroes] the perverter of Peripatetic philosophy."

In the last section of his treatise, Thomas abandons his stance as a philosopher and denounces the un-Christian character of the teaching of some

---

separate et singulares; non autem indiuiduantur ex materia, sed ex hoc ipso quod non sunt nate in alio esse, et per consequens nec participari a multis. Ex quo sequitur quod si aliqua forma nata est participari ab aliquo, ita quod sit actus alicuius materie, illa potest indiuiduari et multiplicari per comparationem ad materiam. Iam autem supra ostensum est quod intellectus est uirtus anime que est actus corporis; in mustis igitur corporibus sunt multe anime, et in multis animabus sunt multe uirtutes intellectuales que uocantur intellectus: nec propter hoc sequitur quod intellectus sit uirtus materialis, ut supra ostensum est." *Ed. cit.*, p 311a.

12 "Sic ergo si intellectus naturaliter esset unus omnium quia non haberet naturalem causam multiplicationis, posset tamen sortiri multiplicationem ex supernaturali causa, nec esset implicatio contradictionis." *Ed. cit.*, p. 311b.

13 "Concedimus autem quod anima humana a corpore separata non habet ultimam perfectionem sue nature, cum sit pars nature humane; nulla enim pars habet omnimodam perfectionem si a toto separetur. Non autem propter hoc frustra est; non enim est humane anime finis mouere corpus, sed intelligere, in quo est sua felicitas, ut Aristotiles probat in X Ethicorum." *Ed. cit.*, p. 313b.

unnamed master of arts. That he is referring to a specific individual seems highly probable, although not absolutely certain. His use of *si quis* is formulaic, as indeed even *tu* would have been, but the specific nature of the matter to which Thomas objects strongly indicates a particular master. It was surely not Siger; it seems more likely to me that it was Boethius of Dacia, because of Thomas's strong emphasis on the proposition that 'this individual man understands.' But a certain identification is nearly impossible, since Thomas makes it clear that he is referring to the master's oral teaching, which has not been put into written form. He concludes with a challenge to his adversary:

> But if anyone boasting of his falsely-named knowledge should wish to say something against these things we have written, let him not speak in corners nor in the presence of boys who do not know how to judge such difficult matters, but let him write against this treatise if he dares, and he will find not only me, who am the least of others, but many others, who are zealous for truth, by whom his error will be opposed or his ignorance remedied."[14]

Although this is a brilliant essay, it is not without its faults. By putting forth a possible, but less probable, interpretation of Aristotle than had Averroes, Thomas has shown that Aristotle's *De anima* is consonant with Christian teaching on the soul, if one grants certain debatable positions, which were however central to Aquinas's thought. As we shall see in the next chapter, even some of those who were in essential agreement with Aquinas felt that he had not made a sufficiently strong case for his position. There was still ample room for legitimate disagreement, and this was not long in making its appearance.

---

[14] "Si quis autem gloriabundus de falsi nominis scientia uelit contra hec que scripsimus aliquid dicere, non loquatur in angulis nec coram pueris qui nesciunt de tam arduis iudicare, sed contra hoc scriptum rescribat, si audet; et inueniet non solum me, qui aliorum sum minimus, sed multos alios ueritatis zelatores, per quod eius errori resistetur, uel ignorantie consuletur." *Ed. cit.*, p. 314b.

CHAPTER EIGHT

REACTIONS TO AQUINAS

Aquinas's challenge was taken up by a number of masters, not all of whom disagreed with his anti-monopsychism stance, in quite different ways. We shall treat five of these here: an anonymous master who was much influenced by Aquinas, especially his commentary on the *Metaphysics*, but who made no direct reference to *De unitate intellectus*; Boethius of Dacia (?), who boldly asserted the rights and independence of philosophy within its proper sphere and claimed that Aquinas's main argument in *De unitate intellectus*, that we have immediate experience of the fact that we ourselves understand, was not proved and ought therefore to be denied; two masters who agreed with much of what Aquinas taught but who criticized his method of presentation, one, Giles of Rome, who felt that none of the arguments yet brought against Averroes was adequate, and an anonymous master who, like Giles, accepted much of Aquinas's position but criticized some of his argumentation; and Siger of Brabant, who began a long philosophical dialogue with Aquinas,[1] during the course of which he modified many of his earlier views, but not all.

ANONYMOUS BAZÁN. 'ANTI-AVERROIST.' Bernardo Bazán has edited a set of questions on *De anima* by an anonymous master,[2] which he dates to 1272-1275. It is apparently the work of a young master and may be characterized as competent but not brilliant. The text we have is a copy of a *reportatio* of an arts course on Aristotle's *De anima* presented *per modum quaestionis*. Its arguments are not fully developed, and quite a few weak arguments on both sides are insufficiently refuted. Its purpose is to interpret the text of *De anima* and occasionally to go beyond it to investigate certain problems arising from

---

[1] There is an extensive bibliography on this, of which the most helpful works are Edward P. Mahoney, "Saint Thomas and Siger of Brabant Revisited," *The Review of Metaphysics* 28 (1974), 531-53; and Fernand Van Steenberghen, *La philosophie au XIIIe siècle*. Philosophes médiévaux 28 (2nd ed., Louvain, 1991), pp. 387-98. See also É. H. Wéber, *La controverse de 1270 à l'université de Paris et son retentissement sur la pensée de S. Thomas d'Aquin*, Bibliothèque Thomiste 40 (Paris, 1970) and the rebuttal by Bernardo Bazán, "Le dialogue philosophique entre Siger de Brabant et Thomas d'Aquin. À propos d'un livre récent de É. H. Wéber, OP," *Revue philosophique de Louvain* 72 (1974), 53-155.

[2] Bernardo Bazán, ed., *Ignoti auctoris Quaestiones super Aristotelis librum De anima* in *Trois commentaires anonymes sur le traité de l'âme d'Aristote* (Louvain, 1971), pp. 351-517.

the text. It is less under the influence of Averroes than many other contemporary products of the arts faculty, but it avoids a number of difficult questions. For example, it does not treat the problem of the relation of faith and reason or note the different methods and subject matter of philosophy and theology, and although it takes some account of the theological attacks on the teaching of the artists, it does not attempt any real accommodation. The master cites with approval works of Albert the Great (properly glossed) and of Aquinas, whose commentary on the *Metaphysics* he uses extensively. However, he shows no knowledge of Thomas's *De unitate intellectus*.

Bazán's discussion of its composition date[3] is learned, thorough, and judicious. He places the work after the end of 1271 (because Aquinas's commentary on the *Metaphysics* was not completed until then and because the author twice refers to book Lambda of the *Metaphysics* as book 12), and before the condemnation of 1277 (since it maintains a doctrine—monopsychism—condemned in that year, and it makes no mention of the condemnation).

Each question begins with arguments for the position to be rejected. The *oppositum* section is, with three exceptions, a citation of Aristotle, which is to be accepted (sometimes with a gloss) and arguments supporting the master's interpretation of Aristotle.

As this master understands Aristotle, the soul is simple *per se*, and he disagrees with his colleague Siger and holds that every form, whether essential or accidental, is simple, although it may be divided *per accidens* by means of matter.[4] The soul is pure form, without any matter (he briefly but definitely rejects universal hylomorphism),[5] but it can nevertheless be individuated. Averroes's claim that the soul cannot be individuated because it lacks matter and because it can grasp universal species is not necessary (i.e., its contrary is not impossible) because: 1- it is free of matter subjectively but not objectively, and this is sufficient for individuation;[6] and 2- because the 'universal species' that the soul grasps is itself particular, "but insofar as such a species is a likeness of several things, it is thus universal, just as the species 'man' is particular and one, and nevertheless with respect to the intellect it is

---

[3] *Ed. cit.*, pp. 366-77.

[4] Lib. 2, q. 38, *ed. cit.*, pp. 460-61.

[5] Lib. 2, q. 4, *ed. cit.*, pp. 419-21.

[6] Lib. 3, q. 21, *ed. cit.*, pp. 510-12.

representative of several men."[7]   This second point is the same as that made
by William of Baglione.

The intellective soul is the substantial form of the body. That through which
the body differs from all other things is its substantial form. Man differs from
all other things through the intellective soul. Therefore the intellective soul is
the substantial form of the body, and neither the opinion of Alexander of
Aphrodisias, who holds that it is the substantial form of the body but is
corruptible, nor of Averroes, who considers it to be separate from the body, is
valid.[8]   When Aristotle says that it is separate and unmixed, he means that it
is separate from and unmixed with any bodily organ.[9]

Because the intellective soul is simple and is the substantial form of the
body, all its powers pertain to the same substance. The human embryo in its
mother's womb does not have a soul, but only 'certain dispositions.'[10] "When
the embryo is developing toward becoming a human being, it first has a
vegetative life and then a sensitive life, and these are intrinsic powers. But
when the intellective soul is then introduced from without, then these
vegetative and sensitive powers are corrupted,"[11] and the intellective soul
itself performs all the vital functions.

The agent and possible intellects are not themselves substances. As Aristotle
explains in *De anima* 3, 5, the soul has something by which all things are
intelligible, and this is the possible intellect; and it also has something by
which the soul understands all things, and this is the agent intellect. Although
they are not substances, they are joined to the intellective soul, which is a
substance.[12]

Averroes's arguments for the uniqueness of the intellect are not conclusive,
and his position is contrary to the Christian faith. The question may be argued

---

[7] "... tamen in quantum ista species est similitudo plurium, sic est universalis: ut haec
species 'homo' est particularis et una, respectu intellectus tamen est repraesentativa plurium
hominum." Lib. 3, q.26, *ed. cit.*, p. 478.

[8] Lib. 3, q. 6, *ed. cit.*, pp. 476-78.

[9] Lib. 3, q. 12, *ed. cit.*, pp. 488-91.

[10] Lib. 2, q. 1, *ed. cit.*, pp. 402-04.

[11] "[Q]uando embryo est dispositus ad hominem, tunc primo vivit vita vegetativa et
deinde vivit vita sensitiva et hae sunt potentiae ab intrinseco; sed quando tunc anima
intellectiva introducitur ab extrinseco, tunc, ea adveniente, potentiae vegetativa et sensitiva
corrumpuntur." Lib. 3, q. 2, *ed. cit.*, p. 470.

[12] Lib. 3, q. 12, *ed. cit.*, pp. 488-91.

either way, and the work ends with two treatments of the question "Whether the intellect is one in number in all men."

In the first,[13] after the introductory *quod sic* arguments, instead of citing Aristotle for the position to be adopted, he says: "Oppositum est fides nostra." He then confirms this by three arguments: 1- that rewards and punishments would in this case be the same for all, regardless of sinfulness or merit, which is impossible; 2- the intellect is the perfection of the body, and the intellect is numbered by the number of bodies, of which there are many; and 3- if the intellect were the same in number for everyone, then when I understand, you must understand, and when I do not, you do not, which is contrary to experience.

Then, denying the position of Averroes, he replies to the questions *quod sic*: 1- to the contention that since the intellect, lacking matter, thereby lacks the principle of individuation, he says that it lacks matter only subjectively, not objectively, and this is sufficient for individuation; 2- to the argument that if the world were eternal, there would now be an actually infinite number of departed souls if the intellect were many instead of one, and this is contrary to what Aristotle says in *Physics* 3, 5, that the infinite cannot exist in act, he replies that Aristotle was here speaking only of the infinitude of sensible bodies, not of separated substances, and therefore, since the soul is a separated substance, there can be an actually infinite multitude of them. Therefore, he says, the arguments of the Commentator are not conclusive.

Our master concludes his course by arguing the same question, this time upholding the position of the Commentator,[14] who posits one intellect in all men and holds it to be separate both with respect to its substance and as a distinct thing, but joined by its operation. This is followed by a summary of Averroes's arguments for his position. Then the master replies to the *quod non* arguments, which he himself had used in the preceding question, thus leaving the matter up in the air.

This master is not an ardent controversialist for either side of the debate. He is aware of the theologians' strictures, and, although he concedes that monopsychism is contrary to faith, he does nothing toward solving the impasse. He seems content to let the ambiguity remain. Although Averroes's arguments are not demonstrative, neither, by implication, are those supporting the faith, since he himself refutes them.

On all matters but monopsychism, his views are much closer to the theologians than were those of Siger and Boethius. His teaching that the soul is the substantial form of the body, and that the intellective soul is simple and

---

[13] Lib. 3, q. 21, *ed. cit.*, pp. 510-12.

[14] Lib. 3, q. 22, *ed. cit.*, pp. 513-14.

performs all the life functions of men were shared by many of the theologians, and several also agreed with him in rejecting universal hylomorphism. Throughout most if the work, he either maintains or implies that each human being has his own soul, although his refutation of Averroes on this point is restricted to showing that his arguments are not necessary; he nowhere claims that the position is demonstrably false.

This is not a work of great merit, but it does provide us with the views of a professor of philosophy around 1271-75 who was certainly not 'Averroist' (except to the extent that, like everyone, he habitually used Averroes as a guide to Aristotle), but on the other hand was not virulently anti-Averroistic.

BOETHIUS OF DACIA (?).   A second response to Thomas's challenge to the 'Averroists' to reply in writing if they dared gave no quarter, and indeed even took the philosophical path further than any other artist, or even Averroes, had, in claiming that it is not true properly speaking that 'this man understands,' or at least it has never been proven. This work is probably a *reportatio* of an arts course *per modum quaestionis* on books 1 and 2 of *De anima*,[15] but it also contains several sections of straightforward commentary.

Its present incomplete state results probably from the fact that the copyist did not finish it, rather than the master's lack of time to complete his course, since there are references in it to matters which are to be treated in book 3. Since this is in some respects a response to Aquinas's *De unitate intellectus* and consequently subsequent to it, and since it was used by Giles of Rome in his *De plurificatione intellectus* (1273-1275), it must have been written between 1270 and 1275. For reasons that I will summarize later, I prefer a date prior to 1272.

Although the work is anonymous, there are good reasons for thinking that its author was Boethius of Dacia. When Giles of Rome was an old man, he reminisced about how, when he was a bachelor, a very famous philosopher had insisted that properly speaking man does not understand. This position is a unique feature of the present treatise. We do not know the names of many great philosophers at Paris during the time that Giles was a bachelor (of arts,

---

[15]   Maurice Giele, "Un commentaire averroïste du traité de l'âme d'Aristote," *Medievalia Philosophica Polonorum* 15 (1971), 1-168 gives much the same text as the same author's *Un commentaire averroìste sur les livres I et II du traité de l'âme (Oxford, Merton College 275, f. 108-121)*, in *Trois commentaires anonymes*, pp. 13-120, but with a less complete apparatus, and including a fuller doctrinal study and bibliography. My references will be to the version printed in *Trois commentaires anomymes*. Giele has also pointed out the close connection between this work and Giles of Rome's *De plurificatione intellectus possibilis* in his "La date d'un commentaire médiéval anonyme et inédit sur le Traité de l'âme d'Aristote (Oxford, Merton College 275, fol. 108r-121v)," *Revue Philosophique de Louvain* 58 (1960), 529-56.

not theology, as Giele very sensibly suggests[16]), i.e., the late 1260s. The author of the work was certainly not Siger of Brabant, since its doctrine is often inconsistent with the latter's *In tertium De anima*. But much about it suggests Boethius: the clear distinction made between the methods of inquiry of the philosopher and the theologian; the vigorous assertion of the rights and competence of philosophy within its proper sphere;[17] and the verbosity and repetitiousness (interspersed with striking aphorisms) of the argumentation; as well as certain phrases that are characteristic of Boethius's style. I shall therefore take the liberty of referring to its author as Boethius. And if we grant that the author was Boethius, then we must place this work prior to *De aeternitate mundi*, which I have elsewhere dated as late 1272,[18] since he had not yet worked out the intricacies of his position on the relation of faith and reason, although the positions of that treatise are implicit in it.

Boethius begins his course by establishing to what extent the science of the soul falls within the domain of the physicist. The physicist may investigate whatever has its being in matter. The soul has its being in matter, except perhaps the intellective soul, about which there is some doubt. But, although the intellective soul may have its *being* separate from matter, the physicist's way concerns the soul's *operation*, in which the soul does communicate with nature and the body. Therefore, there can be natural science concerning the soul, for without matter and physical things, we cannot even prove the existence of the soul. Therefore there is no science of the soul that does not also include the body, and "it is certain that there is no science of the soul separated from the body."[19] Consequently the whole range of arguments that the theologians bring against the artists are irrelevant to a philosophical investigation.

Then he argues, contrary to Siger (in *De anima intellectiva*) and Aquinas, that the intellective soul cannot be the substantial form of the body. To certain people, he says, it appears that the intellective substance gives being to the body, just as the vegetative and sensitive do, so that the soul is universally the perfection and form of the body, but that nevertheless a certain power of the soul, the intellective, is separable. And nevertheless they say and confess that its operation is separate. And they declare that the soul is universally the

---

[16]   *Ed. cit.*, Introduction, p. 18.

[17]   See Paul Wilpert, "Boethius von Dacien—Die Autonomie des Philosophen," *Beiträge zum Berufsbewusstsein des mittelalterlichen Menschen* (Berlin, 1964), 135-52.

[18]   *Medieval Discussions of the Eternity of the World*, p. 153.

[19]   *In Aristotelis libros I et II de anima* I, 3, *ed. cit.*, pp. 26-27.

act of the body because we understand, and the soul is that by which we understand.   But in truth this position absolutely cannot stand, for an operation that is separate from matter and is a separate power does not prove the substance whose being is the form of the body.   This position implies contradictory things, because it says that we understand from the manner in which understanding is a passion in matter, because it says that we understand through the intellect, which is the act of the body.   And these points are contradictory to what was said before, that understanding is a separated passion.

Also, that the operation of understanding is common to the soul and body, or that the soul needs the body for understanding, can be understood in two ways. One way is that understanding is a passion in which the soul needs the body as a subject in which to bring about understanding, as sight needs the eye in seeing—and in this way the soul does not need the body. But the soul does need the body in understanding as an object, not as a subject. In this way, the soul needs the body, for Aristotle thus proves that understanding is not the unique property of the soul.

If the soul were the substantial form of the body, it would necessarily follow that the soul is generable and corruptible, for it would thereby be a material form. "Those who argue against this manner of coupling argue well that we do not understand as a subject of this operation, but as that without which the intellect does not understand."[20]

The climax of this chain of argumentation comes in book 2, question 4. The author has been preparing us for this question, and he spends more time on it than on any other. It is a strict and proper application of philosophical principles and techniques to a difficult problem. Boethius had told us in question 3 that a natural philosopher can only treat the soul as it is joined to the body, since the only evidence we have of its existence is through its operation in the living composite. As Aristotle says, we must progress from things better known to us but posterior in the order of nature to a knowledge of what is prior in nature, though less well known to us. A philosopher can only use natural reason and experience.

The question is: Whether the intellective soul, according to its substance, is joined to the body as its substantial perfection. After giving arguments for and against, Boethius summarizes Aquinas's position from *De unitate intellectus* that the intellect is the act of the body in its being (*esse*) as a form of matter, but is nevertheless a separated power in its operation, refutes it by

---

[20]   "[Q]ui arguunt contra hunc modum copulandi, bene arguunt quod nos non intelligimus sicut subiectum huius operationis, sed sicut illa sine quibus intellectus non intelligit." *Quaestiones de anima* I, 6, *ed. cit.*, p. 59.

argumentation, and shows that it is contrary to what Aristotle says in *De anima* 3, 4 (429b). If it were thus, he says,

> and if the intellect is incorporeal, as Aristotle says and everyone admits, I say that it is not the act of the body as its substance. ... We ought always to look to what is apparent and to conclude from what is apparent that the intellect itself is not the form of the body. ... [Aristotle teaches that] the intellect needs the phantasms supplied by the senses in order to understand anything, but it needs the body as an object, not as a subject, and so understanding is proper to the soul.[21]
>
> We say that the soul (i.e., the intellective) is not the act and form of the body. But a great absurdity seems to follow from this, because if it is so, then it follows that we do not understand, and from this it follows that man does not understand. Nor does the position of the Commentator, that because the objects of knowledge are coupled to us, therefore the intellect is too, have any force.[22]

Then follows a long string of arguments against the contrary position, including Averroes's (pp. 72-75)—this from an author who is labelled an Averroist—ending with the words: "This speech is long, but it has a short solution."[23]

The solution is one of the most audacious bits of philosophical reasoning of the Middle Ages, and it attacks precisely the point that had played so crucial a role in Aquinas's *De unitate intellectus* and was accepted by Siger without investigation in *De anima intellectiva*, namely the subjective experience the individual has of understanding. But Boethius will not accept the common

---

[21] ["Si est ita, scilicet secundum quod intelligere non est passio in materia, sed intellecta sunt separata,] et si intellectus est incorporalis, ut vult Aristoteles et sicut omnes confitentur, dico quod non est actus corporis ut sui substantia ... Semper enim debemus ad apparentia respicere et ex apparentibus concludere hoc, scilicet ipsum <intellectum> non esse formam corporis; ... communicat enim corpori et eget corpore sicut obiecto, non sicut subiecto. Ita quod intelligere proprium est animae et non corpori sicut subiecto." II, 4, *ed. cit.*, pp. 71-72.

[22] "Dictum est nunc animam esse non actum et formam corporis, scilicet intellectivam. Sed videtur inconveniens ex hoc sequi maximum, quia, si ita sit, tunc sequitur quod nos non intelligimus; ita quod sequitur ex hoc quod homo non intelligit. Nec valet via Commentatoris: dicit quod quia intellecta nobis sunt copulata, ideo et intellectus nobis copulatus est." II, 4, *ed. cit.*, p. 72.

[23] "Iste sermo longus est et habet brevem solutionem." II, 4, *ed. cit.*, p. 75.

explanation of this experience. "If some illogical proposition is only supposed, admitted, or conceded," he says,

> it is easy to conclude many illogical things. Some people accept the fact that this man, properly speaking, understands, but they do not prove it. They simply argue from its supposition and do not ask whether it is true. Therefore, I do not concede that a man, properly speaking, understands. If this is conceded, I do not know how to respond. But I deny it, and rightly so, and so I shall respond easily.
>
> Aristotle says in *De anima* 3, and so do his adversaries, that it follows from the fact that understanding does not have an organ and is separate from matter, that properly speaking man does not understand in the same way that he senses. You say: "Aristotle says in *De anima* 1 that it is man and not the intellect that understands." One ought not to ask: "Does man understand, and not only the intellect," except in the manner in which understanding is common to the soul and the body, and not only the soul. But now Aristotle seems to determine in book 1 that understanding is not the exclusive property of the soul, but of the body and the soul; and the way in which understanding is common to the body is that it does not occur without phantasms. But this does not mean that understanding is the perfection of man, but rather that it needs man as an object. Understanding is not in matter as sight is in the eye, and consequently the intellect is not as man's perfection, but as separate from matter. It needs the material body as an object, not as its subject, and to that extent one can say that man understands; but this is not in the same way as man senses.
>
> If you say that man, properly speaking, understands, [I reply that] it is not proven, and therefore it ought to be denied.[24]

---

[24] "... nam aliquo inconvenienti dato, et illo non probato ibi, alibi supposito, confesso et concesso, multa inconvenientia facile est concludere. Isti autem accipiunt quod homo proprie intelligit, nec hoc probant. Ex hoc supposito arguunt. Quodsi istud suppositum non est verum, non arguunt. Unde, quod homo proprio sermone intelligit, non concedo: illo tamen concesso, nescio respondere; sed istud nego et merito; ideo faciliter respondebo. Si enim, ut vult Aristoteles *tertio huius* et adversarii etiam, si intelligere organum non habet, imo est abstractum a materia, ex hoc sequitur ut proprio sermone homo non intelligat sicut sentit. Dicis: Aristoteles dicit hominem et non intellectum intelligere, *primo huius*. Non est quaerendum hominem intelligere et non intellectum tantum, nisi ex hoc modo quo intelligere est commune animae et corpori, et non animae tantum. Nunc autem Aristoteles videtur determinare *primo huius* quod intelligere non est proprium animae, sed animae et corpori; et

And after belaboring the point in several more arguments, Boethius concludes with a direct attack on the assertion that we experience ourselves to understand:

> If you say: "I experience myself to understand," I say that this is false. Rather, the intellect, united to you naturally and the mover of your body, is what experiences this, just as also the separate intellect experiences the objects of knowledge to be in itself. If you say: "I, the aggregate of body and intellect, experience myself to understand," this is false. Rather, the intellect, needing your body as an object, experiences this, communicating it to the aggregate in the way we have mentioned.[25]

It is understandable why Giles of Rome and Anonymous Van Steenberghen complained that there was no disputing with such people..

Boethius's position is that since a natural philosopher can only investigate the soul insofar as its operations are manifested through the body—for this is the only natural knowledge men have of it—that it is necessary for a philosopher to draw inferences from his experience to their logical conclusions. To the extent that philosophy can tell us anything about the intellect, we must say that it is 1- immaterial; and 2- unique. The individual's experience of understanding can be explained satisfactorily on this basis, and it is not necessary, or even permissible, to employ an unproved proposition (such as 'this man understands') in an investigation of the soul's nature.

GILES OF ROME. At this point in the controversy, the young Augustinian Hermit, Giles of Rome, composed a purely philosophical treatise in the form

---

modus per quem est commune corpori quoniam non est sine phantasmate. Hoc autem non est ut intelligere sit perfectio hominis, sed eget homine ut subiecto. Sic non est dicere intellectum intelligere, sed hominem, non ex hoc modo quo intelligere sit in materia, ut videre in oculo, et per consequens non ut perfectio, sed ut separatum a materia. Eget tamen materiali corpore ut obiecto, non ut subiecto suo; et pro tanto est dicere hominem intelligere; tamen non est ita ut dicimus hominem sentire. Si dicas quod proprie homini <convenit intelligere>, non est probatum, et ideo hoc est negandum." II, 4, *ed. cit.*, p. 75.

[25] "Tu dices: ego experior et percipio me intelligere, dico quod falsum est; imo intellectus unitus tibi naturaliter, sicut motor tui corporis et regulans, ipse est qui hoc experitur, sicut et intellectus separatus experitur intellecta in se esse. Si dicas: ego aggregatum ex corpore et intellectu experior me intelligere, falsum est; imo intellectus egens tuo corpore ut obiecto experitur hoc, communicans illud aggregato dicto modo." II, 4, *ed. cit.*, p. 76.

of a *quaestio*,[26] without any display of rancor, stating the crux of the dispute, summarizing the arguments given by Averroes, his supporters and opponents, evaluating them, and concluding with what he felt to be a better refutation of Averroes than any of his predecessors had managed. He divided his treatise into four parts: first, the arguments of Averroes in behalf of his own position, those he gives against his own position, and his replies to them; second, a clear statement of Averroes's position; third, a summary of arguments against Averroes's position, "which do not disprove his position as they should;" and fourth, an analysis and refutation of Averroes's view on the uniqueness of the intellect.

The first two parts, summarizing the text of Averroes, are admirably clear and fair and conclude with the statement that "the position of the Commentator seems in some manner to be rational."[27] This is followed by an evaluation of opposing arguments, several of them taken from Aquinas's *De unitate intellectus*. The first and most basic of these is the contention that, on Averroes's view, it would not be true that man understands, but rather that he is understood. Aquinas's discussion of what 'man' is is also presented, as is the analysis of the manner of the connection or separation of the individual man and the intellect, which holds that according to Averroes, the intellect is rather separated from us than joined to us.

"But one must note the fact," he concludes," that all the preceding arguments conclude nothing unless one supposes that man, properly speaking, understands. ... One must declare therefore that all these arguments either lead to the illogical position that man does not understand ... or from the supposition that man does not understand an absurdity follows."[28]

Then, turning to the treatise of Boethius of Dacia (?), Giles points out that some people, in order to avoid the force of these arguments, concede that man, properly speaking, does not understand. With this concession, the aforementioned arguments are solved by Aristotle's dictum in *Physics* 1 that if one illogical thing be granted, many others follow. And those who argue

---

[26] This work is printed under the title *De intellectu possibili contra Averroim questio aurea* in Aegidius Romanus, *Super librum de Anima, De materia celi, De intellectu possibili, De gradibus formarum* (Venice, 1500, repr. Frankfurt a. M., 1982), fols. 91ra-94rb. The author himself, in the last line of the treatise, entitles it "De plurificatione intellectus possibilis."

[27] "apparet positionem Commentatoris aliquo modo esse rationabilem." *De intellectu possibili, ed. cit.*, fol. 92rb.

[28] "Advertendum quod omnes rationes pretacte nihil concludunt nisi hoc supposito quod homo proprie acceptus intelligat. ... Declarandum est ergo quod omnes ille rationes vel ducunt ad inconveniens illud scilicet quod homo non intelligit ... vel ex ista suppositione scilicet quod homo non intelligit ducunt ad illa inconvenientia." *Ibid., ed. cit.*, fol. 92vb.

that it is true that man understands are deficient in three respects: first, because they contradict themselves; second, because they posit something by whose concession no one can argue with them; and third, because they make an argument by which they mean to defend Averroes's contention that man does understand, but they do not retain the position of the Commentator. This is followed by a summary and amplification of these points as in Boethius of Dacia (?).

The conclusions of all these arguments, Giles continues, are true to some extent, but the argumentation is faulty. The whole force of an argument, he says, is contained in the premisses, and if we inquire diligently into how much force the premisses contain, it becomes clear that they conclude nothing against the Commentator.[29]

In the fourth part, Giles says that it remains to present those arguments by which the position of the Commentator is shown to be untenable, and consequently to disprove the subsidiary arguments that attempt to save his position, as well as to show that the position of the Commentator is not only false, but impossible. Giles's strategy is to show that six absurdities follow from Averroes's position: 1- that man does not understand; 2- that the intellect, in understanding, does not need the body; 3- that the intellect is not able to receive any species; 4- that the agent intellect does not bring about the understanding of all things ("quod intellectus agens non est omnia facere"); 5- that the possible intellect understands itself by itself and not by understanding other things; and 6- that our possible intellect is not in the genus of intelligible things as prime matter is in the genus of beings. "All of these things," he says, "are against the basic principles of the Peripatetics and against truth itself."[30] Then, after providing proofs for his contention that these un-Peripatetic, irrational, and impossible consequences follow from Averroes's position, he says that the Commentator himself realized that his position could not stand on rational grounds alone, and this is why he spoke in the manner of an imprecation or persuasion when he wrote: "And thus I ask the brothers who see this work to write down their doubts, and perhaps in this way the truth will be found, if I have not yet found it. And if I have found it, as I think, then it will be made more plain by these questions."[31]

---

[29] *Ibid., ed. cit.*, fol. 93rb.

[30] "Que omnia sunt contra fundamenta peripateticorum et contra ipsam veritatem." *Ibid., ed. cit.*, fol. 93ra.

[31] "propter quod ipsemet videbat quod positio eius stare non poterat: ideo quando debebat positionem suam ponere inceperit precari et modo persuasive loqui et credebat quod si eius ratio non prodesset eius persuasio hoc supplicaretur unde dicebat: Rogo fratres," et cetera. *Ibid., ed. cit.*, fol. 94ra.

It now remains to reply to Averroes's arguments in a compelling manner. To accomplish this, Giles holds that they are all, in fact, one argument, "for they all proceed from this root, that the intellect is immaterial and therefore is not multiplied,[32] and he goes on to show how each of the other five arguments can all be reduced to this one. The way to answer Averroes therefore is to show that the fact that the intellect is immaterial is not inconsistent with its being multiplied, and he argues this point much more fully than did Anonymous Bazán. Giles dismisses hylomorphism without even mentioning it in his explanation of what is meant by the term 'material form.' Forms are not called material, he says, by reason of a matter from which they are, but by reason of the matter in which they exist, for no form has matter as part of itself."[33] But some forms are the perfection of matter and some are not; the former are called material forms, the latter immaterial forms. Some material forms are educed from the potency of matter according to both their essence and their disposition; others according to their disposition only and not according to their essence. The only form of the latter kind is the intellective soul. Among the forms that are perfections of matter, the only one not educed from the potency of matter according to its essence is the intellective soul, which is so educed only according to its disposition. Therefore the intellective soul is midway between forms completely separated and forms immersed in matter, and since it holds a middle position between these two, thus the powers that follow upon the soul hold a middle way. Just as all the powers that are in separated substances are not perfections of any matter, and the powers that follow upon other material forms, such as the souls of beasts, which are all material and organic powers, thus also the powers that follow upon our soul hold a middle position between material forms and separated substances, because sometimes the powers of the soul are organic and sometimes they are not. Since the intellect is a power following the manner of the essence of our form, or our soul, as in a subject, it will be founded in the very essence of our soul and not in some corporeal organ. Therefore it is necessary that the intellective powers, or the intellects on which they are founded, be multiplied as a subject.

"Thus it is clear that those two propositions are not inconsistent with each other: that the intellect is immaterial (that is, it is not an organic power), and that it be multiplied by the multiplication of individual men. ... The Philosopher, wishing to designate the fact that the intellect is founded in the

---

[32] "omnes enim procedunt ex ista radice quia intellectus est immaterialis ideo non plurificatur." *Ibid., ed. cit.*, fol. 94ra.

[33] "forme dicuntur immateriales non ratione materie ex qua sed ratione materie in qua: nam nulla forma habet materiam parte sui, vel materia ex qua." *Ibid., ed. cit.*, fol. 94rb.

essence of the soul as in a subject, but the senses are founded in an organ, said that the place of visible species is the eye, in which is founded the power of sight as in a subject; ... but the place of intelligible species is the intellective soul, since in it is founded the intellect, as in a subject."[34] Then on the basis of this conclusion he replies to each of the six arguments of Averroes.

Giles has concentrated on two of the most vexing problems concerning the intellect, as the argument stood in the mid-1270s: how the intellect could be multiplied if it were wholly immaterial; and the way it is true that man understands. Concerning the first, he eschewed the path of hylomorphism, which had been adopted by the followers of Bonaventure, especially William of Baglione and John Pecham, and had instead attempted a clearer presentation of Aquinas's position, but without the formula that the soul is individuated *ex corpore*. Concerning the second, he attempted to sever the dependence of the position of the intellect's multiplicity from the question of whether it is true that man, strictly speaking, understands. He continued to consider Boethius of Dacia's (?) position absurd ("inconveniens"), but he shifted the crux of the anti-Averroist argument from this point to the lack of a contradiction between the soul's immateriality and its multiplicity. His treatment added nothing philosophically to Aquinas's, but it did present its main points in a more acceptable manner: the theologians would not be offended by the phrase that the soul is individuated by virtue of the body, and the philosophers could ignore the problems involved in what it means to say that man understands, since, in Giles's view, this was not central to the discussion.

ANONYMOUS VAN STEENBERGHEN. There us an extremely interesting set of questions on the soul that has been known for a long time, strikingly similar to Giles's question but considerably superior to it. In 1931 Fernand Van Steenberghen edited this work and, following Grabmann, he attributed it to Siger of Brabant.[35] This attribution met with widespread criticism, and in 1971 Van Steenberghen re-edited the work and this time, correctly I think,

---

[34] "et sic apparet quod illa duo sibi invicem non repugnant scilicet quod intellectus sit quid immateriale hoc est non sit virtus organica et quod multiplicetur multiplicatione individuorum hominum... hoc autem volens designare philosophus quod intellectus in essentia anime fundatur ut in subiecto: sensus vero in organo: ait quod locus specierum visibilium est oculus in quo fundatur virtus visiva sicut in subiecto ... sed locus specierum intelligibilium est essentia anime intellective cum in ea fundetur intellectus ut in subiecto." *Ibid., ed. cit.*, fol. 94rb-va.

[35] Fernand Van Steenberghen, *Siger de Brabant d'après ses oeuvres inédites* 1. Philosophes Belges. Textes et Études 12 (Louvain, 1931), pp. 11-156.

considered its author anonymous.[36]   Although there is some similarity to the authenticated works of Siger, there are also considerable differences. If these are accounted for as representing developments in Siger's thought, this work would have to be placed after the questions on *De causis*, which were composed in 1276; and although this is not completely impossible, it is not likely. I shall therefore follow Van Steenberghen's reconsidered judgment and consider the work to be by an anonymous master of arts at Paris. It is clearly posterior to Aquinas's *De unitate intellectus* (1269-70) and Boethius of Dacia's (?) questions on *De anima*. Since it cites Aquinas's commentary on the *Metaphysics*, it must also be posterior to 1272. And the collection of works in which it is contained was put together before 1277.   Van Steenberghen's dating of 1273-1277 is sound.

The author takes a very independent attitude toward his authorities and his contemporaries. He dismisses a point of Aristotle's by saying that Aristotle wrote his *Categories* when he was young and perhaps made a mistake. He criticizes Averroes for not understanding Aristotle correctly. He criticizes Boethius of Dacia (?) for contradicting the Commentator (on whether man understands). And, although his own position is very close to that of Aquinas, he on several occasions takes Thomas to task for employing faulty argumentation.

This anonymous master insists on the unity of substantial form. Soul and body are related as matter and form, from which is made one thing. The union is immediate, and Averroes is incorrect in saying that they are only related through the phantasms or through operations. Since the soul is the first act of the body, the body cannot exist in act apart from the soul; if we consider the body apart from the soul, it will not be a body in act, but only in potency. It is the soul by which matter is constituted as part of a substance (*substantiatur*), disposed into three dimensions, and organized. Its subject is prime matter, not a previously organized body.

In this respect, the author goes beyond Aquinas in accounting for the development of the embryo. There is one essence of the soul, in which the vegetative, sensitive, and intellective are rooted. The vegetative does not precede the sensitive, nor does the sensitive precede the intellective. The vegetative and sensitive powers are not in the essence of the soul as acts, but as potencies. The intellective soul is the first act of the body, and the other two powers are its potencies. The master considers the view of Aquinas on the development of the embryo and rejects it.

---

[36]   Fernand Van Steenberghen, ed., *Un commentaire semi-averroïste du traité de l'âme* in *Trois commentaires anonymes sur le traité de l'âme d'Aristote*. Philosophes médiévaux 11 (Louvain, 1971), pp. 121-348. My references are to this edition.

> In another way, one might say that man is generated by many generations. First the semen must be generated; then the semen is corrupted and something else, the vegetative, is generated through the corruption of the form of the semen. Then that is corrupted and something else, the sensitive, is generated, and nevertheless the vegetative remains with respect to its powers; whence the corruption of one thing is the generation of another. And finally, when the sensitive is in the highest degree disposed toward the intellective, then the sensitive is corrupted and the intellective is generated. To this argument, I concede the major premiss. But to the minor, I say that the vegetative in man is not generated prior to the intellective, but afterwards. And one must say the same thing about the sensitive.[37]

Indeed, "the vegetative and sensitive in man are not induced from the potency of matter by a material agent; rather the whole soul of man, which is a simple form that is rationative, is educed from without. But its powers are vegetative and sensitive."[38] There is no other form in man than the intellective, which contains as its powers the nutritive and the sensitive. The sensitive therefore is separable, and the principle of its operation remains in the power of the separated form, which still has the aptitude of sensing if it should be in a body. Since the vegetative and sensitive are powers in an organ, when the organ is corrupted they can be corrupted *per accidens* but not *per se*.

This is the only thirteenth-century master I have found, aside from Robert Grosseteste, who considers the whole soul to be infused at conception. It is probably to this work, rather than to Grosseteste's, that Matthew of Aquasparta refers in his refutation of this position.[39]

---

[37] "Alio modo convenit dicere quod homo generatur multis generationibus. Primo enim oportet generari semen; postea corrumpitur semen et generatur aliud, ut vegetativum, per corruptionem formae seminis; deinde illud corrumpitur et generatur aliud, ut sensitivum, et remanet tamen vegetativum virtute: unde corruptio unius est generatio alterius; ulterius, cum summe disponitur sensitivum ad intellectivum, tunc corrumpitur sensitivum et generatur intellectivum. Ad rationem allatam concedo maiorem; ad minorem dico quod vegetativum in homine non generatur prius quam intellectivum, sed posterius; similiter dicendum de sensitivo." *Quaestiones in De anima* 2, q. 7, *ed. cit.*, p. 208.

[38] "... vegetativum et sensitivum in homine non inducantur de potentia materiae per agens materiale: imo tota anima hominis, quae est forma simplex, quae est rationativa, educitur ab extrinseco; istius tamen virtutes sunt vegetativum et sensitivum." *Loc. cit.*

[39] See below, ch. 9, p. 165.

Up to this point, the master has been emphasizing the soul's function as a form of matter, the body, to which it gives being. He must now consider how it can be such and still not be a purely material form that must perish when the composite perishes. He does this by distinguishing two ways in which a form can be the act of a composite. In one way, it can be totally immersed in matter, both constituted in being through matter and educed from the potency of matter, as are all material forms except for the intellective soul. In another way, a form can be the act of a composite not because it is immersed in matter, but because it has being *per se*, which it communicates to matter. It is in this second way that the intellect is the form of the composite.[40]

The master spends some time showing the absurdities that follow from the assumption that the soul contains matter, whether a kind of matter subject to substantiality only (as some people claim for the soul), to substantiality and corporeity, or to substantiality, corporeity and corruptibility. There is in the soul, however, something material and something formal: material, such as the possible intellect, "quo est omnia fieri," because the soul, through the possible intellect, is in potency to all material forms; formal, such as the agent intellect, "quo est omnia facere." But the material principle in the soul differs from prime matter, because the latter is the principle of corruption, the former is not. They also differ in the method of reception, since prime matter receives individual forms, while the intellect receives material forms universally.[41]

In discussing whether the intellect is a material power, the master notes that Averroes holds that the intellect is not joined in being to the body and consequently is not its perfection; and he summarizes Aquinas's rejoinder (attributed) from *De unitate intellectus*, which uses the example of a heavy thing's being accidentally prevented from being down. "But," he says, "this does not solve the question, because a heavy thing is up violently, and no case of violence is eternal."[42] His own solution is that the human soul holds a middle place between forms that are completely material and those that are completely immaterial.

When the master comes to discuss the issue of monopsychism, he finds himself in a difficult position: he must proceed as a philosopher, but he is

---

[40]   *Quaestiones in De anima* 3, q. 1, *ed. cit.*, p. 302.

[41]   *Ibid.*, 3, q. 1, *ed. cit.*, pp. 304-05.

[42]   The whole passage is as follows: "[A]d rationem Commentatoris respondet Thomas de Aquino quod verum est quod inest alicui per essentiam suam, non potet ei inesse per aliquid quod est de essentia eius; potest tamen per aliquid accidens ei, ut patet de gravi: per naturam suam natum est esse deorsum, tamen per accidens potest esse sursum. Sed illud non solvit, quia grave est sursum violenter, et nullum violentum aeternum."

reluctant to adopt a position that is contrary to faith. But he does not think that either Aquinas or Boethius of Dacia has treated the problem adequately. First he examines the *fundamentum* if the Commentator, which, as he states it, is that no separated form can be multiplied. He states at the outset that it is contrary to faith to say that an immaterial form cannot be multiplied numerically in one species, but that according to Aristotle and all of the philosophers, such multiplication is impossible, because a form *per se* is indivisible and is divided only insofar as it is received in matter, since every multiplication is through matter. But even when form is numbered as it is in matter, as it is understood it is one. And so there is one intellect for all men, as they are understood in species. And he concludes that these arguments of the Commentator seem to be unanswerable, but the opposite is true by faith.[43]

He then gives a concise and accurate account of Averroes's arguments, the arguments that Averroes raises against his own position, and his replies to them, and then says that although Averroes's position is probable, it is not true. And in a manner similar to that of Siger, he says that we should therefore bring arguments against it. First he borrows but recasts two arguments from Aquinas's *De unitate intellectus*, one asserting that from Averroes's position it follows that man does not understand but rather is understood, "because we do not say that a house builds, but that it is built;"[44] and the other, which Siger had also found convincing, that no man now existing can understand, because we only understand because the species that are in the phantasms are received in the possible intellect, and if there were only one intellect and the human race were eternal, it would be full of forms.[45] Next he considers the position maintained by the master I have assumed to be Boethius of Dacia. Some people, he says, seem to evade such arguments by conceding the conclusion (i.e., that man does not understand). They say that properly speaking the intellect understands, and man does not understand in the same way that a sense senses; and the intellect is united to us in the same way as the Intelligences that move the heavenly spheres. Our master criticizes those who solve the problem this way on three grounds, as had Giles: first, they contradict themselves; second, they deny that the terms used by their opponents signify anything, and, as Aristotle says in *Metaphysics* 4, there is no disputing with them; and third, they do not hold to

---

[43] "Et istae rationes sunt Commentatoris, quasi indissolubiles; tamen oppositum verum est per fidem." *Ibid.*, 3, q. 6, *ed. cit.*, p. 312.

[44] This argument is borrowed from Giles of Rome, *De plurificatione intellectus possibilis*. See Giele, "La date," p. 535.

[45] *Ibid.*, 3, q. 7, *ed. cit.*, p. 316.

the path of the Commentator.   They contradict themselves because they
concede that man perceives the understood thing to be united to themselves,
and this could not be except by understanding through the intellect.   They do
not hold to the path of the Commentator because, although he posits the
possible intellect to be a separate substance according to its being, he
nevertheless does not posit that the intellect understands, but that man
understands through the intellect.[46]

Finally, although Averroes's *fundamentum*—that a separate immaterial form
cannot be multiplied—cannot be denied according to philosophy,[47] it can be
interpreted in such a way that the Commentator's conclusions need not follow
from it. For to the assertion that the intellect is immaterial, one can reply that
'immaterial' can be understood in two ways: either because something does
not have matter as a part of itself; or because it is not the act of matter. In the
first way, the intellect is immaterial, but in the second way it is not. It is not a
material act in the same way that material forms are, because it occupies a
middle position between separated substances and completely material forms.
 Aristotle says that the intellect is a power of the soul, and so 'intellect' can be
understood for 'soul' or for 'a power of the soul.' This solution seems to be
much in line with that of Aquinas, although differently phrased: the soul,
although it does not contain matter as a part of itself, is nevertheless the act of
matter, while still having a power, intellection, that is not the act of matter.
And so, while Averroes's *fundamentum* cannot be denied by philosophical
means, still the Christian doctrine that each human soul is individual can be
saved.

In questions 16 through 18 of book 3, the author investigates the way the
intellective powers are related to the soul as a whole. Both the agent and
possible intellects are powers of the same substance of the soul.[48]   An
operation is not to be attributed to a power but to the substance that possesses
the power; for example, we do not say that sight sees or that understanding
understands.[49] Therefore one should not say that the agent intellect abstracts,
but that the intellective soul does so by means of the agent intellect. Nor
should we attribute the reception of the phantasms to the possible intellect, but
to the intellective soul by means of the possible intellect. The operations of

---

[46] *Ibid.* 3, q. 7, *ed. cit.*, p. 317.

[47] "Fundamentum eius non potest negari secundum philosophiam." *Loc. cit.*

[48] *Ibid.*, 3, q. 17, *ed. cit.*, p. 334.

[49] *Ibid.* 3, q. 18, *ed. cit.*, p. 335.

receiving and abstracting make one operation of understanding, because action and passion are one motion, and one motion is terminated at one end.[50]

This work is an excellent example both of the impact of Aquinas's thought on the arts faculty and of the lack of uniformity (despite obvious similarities) among those masters who might be classified as 'pure' or 'independent philosophers.' Although this master, like Siger, is concerned to find concord between philosophy and faith, he is not willing to accomplish this by other than philosophical means. Unlike Siger's later works, we find here no suggestion that philosophy is an inadequate instrument, only that it must be better employed.

SIGER OF BRABANT. The most extended response to Aquinas was that of Siger of Brabant. Although the existence of two other works he may have written has been inferred from citations in later works, I find it sufficiently difficult to divine a man's true doctrine from well edited texts of works which we possess, and I shall therefore ignore those we do not. Siger took seriously those points of Aquinas that he considered valid, and attempted to refute those he did not.

DE ANIMA INTELLECTIVA.[51]   De anima intellectiva is a serious and thoughtful work, in which Siger takes account of Aquinas's De unitate intellectus, modifies some of his earlier positions as a result of it, holds firm to others, and identifies several areas of uncertainty. He makes it very clear several times that his purpose in this treatise is limited to expounding Aristotle's meaning and to discovering as much of the truth as can be obtained by human reasoning and sense experience. He specifically states that in difficult matters where the resources of philosophy are inadequate to the problem, one must hold to the teaching of faith, which is superior to all human reason.

Aquinas, in De unitate intellectus, had severely criticized the Averroist position as being a perversion of Aristotle's thought and those who upheld it as being ignorant of certain Peripatetic commentators, especially Themistius. Consequently Siger's strategy in De anima intellectiva is to stick close to Aristotle's text and to employ the commentary of Themistius extensively. He is primarily concerned, however, not to carry on an argument with Aquinas, but rather to discover the truth about Aristotle's teaching on the soul. He points out several places where Aristotle's words are not clear, others where they seem to be contradictory, and still others where Aristotle did not say

---

[50]   Loc. cit.

[51]   Edited by Bernardo Bazán, Siger de Brabant, Quaestiones in tertium De anima, De anima intellectiva, De aeternitate mundi, (Louvain/Paris, 1972).

anything (such as the state of the intellective soul separated from the body) because it lay outside the bounds of human reason and experience.

The most interesting and important chapter of *De anima intellectiva* is chapter 7, "Utrum anima intellectiva multiplicatur multiplicatione corporum humanorum," for it is on one's interpretation of this chapter that the doctrine of the entire treatise depends. Siger is still under the spell of the Averroistic interpretation of Aristotle's doctrine on the uniqueness of the intellect, but he is beginning to have serious philosophical doubts and is cognizant of the power of several arguments for the other side.

He begins this chapter by saying that this question must be diligently considered insofar as it is within the province of philosophy and as it can be understood by human reason and experience. He specifies that he is seeking the meaning of the philosophers rather than the truth and that he is proceding philosophically, "for it is certain that according to Truth, which cannot lie, intellective souls are multiplied by the multiplication of human bodies. Nevertheless, some philosophers thought the contrary."[52]   First he presents the arguments that seem to establish the contrary. Then he adds: "But there are also very difficult arguments by which it is necessary that the intellective soul is multiplied by the multiplication of human bodies, and there are also authorities for this position,"[53] all taken from Aquinas's *De unitate intellectus*, the most telling of which is the last, that Aristotle says "the intellect is in potency to intelligible species and the reception of species, and is devoid of species. But if it were unique, it would always be full of species and the agent intellect would be destroyed."[54]   He does not attempt to determine the question here but confesses that because of its difficulty he has long been in doubt as to what the philosophical solution of the question should be and what Aristotle actually meant concerning it. He then reiterates the position he had stated at the beginning of the chapter, that when there is such a doubt, one must adhere to the teaching of faith, which is superior to all human reason.

---

[52]   "Certum est enim secundum veritatem quae mentiri non potest, quod animae intellectivae multiplicantur multiplicatione corporum humanorum.   Tamen aliqui philosophi contrarium senserunt." *De anima intellectiva*, cap. 7, *ed. cit.*, p. 101.

[53]   "Sed et sunt rationes multum difficiles quibus necesse sit animam intellectivam multiplicatione corporum humanorum multiplicari, et etiam ad hoc sunt auctoritates." Cap. 7, *ed. cit.*, p. 107.

[54]   "Philosophus vult intellectum esse in potentia ad species intelligibiles et receptionem specierum, et denudatum a speciebus; quod si sit unus, erit semper plenus speciebus et destruetur intellectus agens." Cap. 7, *ed. cit.*, p. 108.

The stance taken by Siger in this chapter makes it difficult to interpret much of the rest of this work. One would like to be certain in each case whether Siger is considering the intellect to be proper to each human being or unique to the species. But the treatise is consistently ambiguous on this point. Edward Mahoney considers him to have maintained the uniqueness of the intellect throughout the work,[55] and this is certainly a legitimate interpretation. But it is equally possible to understand him as admitting that each man has his own intellect, and that it is separate from the body in the sense of not operating through a bodily organ and united to it because without the phantasms provided by the body the intellect knows nothing. But there are also difficulties with this interpretation, and the most likely explanation is that Siger himself was unsure, and hence ambiguous, on this crucial point.

Siger begins the treatise by reversing himself on one of the major points of his argument in *In tertium De anima*, and asserting that the soul is the form, or act, of the body, as a substance not an accident, giving being to the body and making one thing with it, not separate from it in being. It is the intellect that is thus the form of man, since man is man through his intellect, which would not be the case if the intellect were not the form of man. But the intellective soul is a different sort of thing from the vegetative and sensitive souls, since it does not operate through a bodily organ and does not communicate with matter. It is impassible and is potency without matter and is therefore separate in its being from matter and the body. It is this that enables it to remain in existence after the corruption of the body.

But unless the intellective soul, which we know only through its operation, viz. understanding, were united to matter in some way, it would not be true to say that man understands. "It is separated from matter in one way and united to it in another. Whence Aristotle sometimes says that the intellect is separate from the body and does not communicate in matter with intelligibles, and sometimes says that it is the act of the body and is conjoined to magnitude."[56]

At this point, he disagrees with Albert the Great and Aquinas by name, criticizing both men for not applying themselves exclusively to discovering Aristotle's meaning.[57] His quarrel with Albert is that the latter had asserted

---

[55]   Edward P. Mahoney, "Saint Thomas and Siger of Brabant Revisited," *Review of Metaphysics* 28 (1974), 531-53, on pp. 541-42.

[56]   "Anima igitur intellectiva aliquo modo est unita corpori et aliquo modo separata ab eo. Unde et ipse Philosophus aliquando dicit intellectum separatum a corpore et communicare in materia cum intelligibilibus, aliquando dicit eum esse actum corporis et coniunctum magnitudini." Cap. 3, *ed. cit.*, pp. 80-81.

[57]   "Isti viri deficiunt ab intentione Philosophi, nec intentum determinant." Cap. 3, *ed. cit.*, p. 82.

that the sensitive and vegetative souls pertained to the same substance as the
intellective soul, which is therefore, like them, multiplied with the
multiplication of bodies. Siger denies here, and more fully in chapter 8, that
the intellective soul pertains to the same substance as the vegetative and
sensitive. His principal disagreement with Aquinas is that he thinks Thomas's
view of the nature of the union of soul and body, namely that the soul is
united to the body in its substance as the body's form but separate from it in
its operation of understanding, would necessitate attributing understanding to
matter rather than to man, since the operation could not be separate from the
substance. Siger holds that the operation of the intellect cannot be separated
from its substance and that the intellect cannot be involved in matter in any
way because then one would say that the body understands. But the intellect
uses no bodily organ. Still, it needs the phantasms supplied by the senses in
order to know anything, and it is in this operational sense only that it is united
with the body—the relationship is like that of a pilot to a ship, not like that of
a seal to wax. It is nevertheless properly called the form of man because it
supplies the proper differentia of man, i.e., rational, which constitutes the
form and perfection of man. In trying to clarify this, he says that the
intellective soul is separate from the body and can be called its form because
it is an *intrinsicum operans*, or internally operating agent, with respect to the
body, since the operation of the *intrinsicum operans* denominates the whole.
But he admits to some uncertainty:

> Whence, through those things which the Philosopher says
> about the soul in general (*in communi*), it is not known how
> the intellective soul is the act or form of the body. ... But in
> those places where he is investigating the proper nature and
> *ratio* of the intellect, as in book 3 of this same work, he
> plainly means that it does not communicate in matter with
> other things, and is separate from the body, and does not
> have an organ, and is impassible, and is potency without
> matter. And all the aforesaid he shows not only of the
> power, but also of the substance of the intellective soul.
> We say that the Philosopher taught this concerning the
> union of the intellective soul to the body; but we prefer the
> teaching of the holy Catholic faith *if it should be contrary
> to the teaching of the Philosopher*, as in all things
> whatsoever.[58]

---

[58] "Unde Philosophus innuit quod definitio animae in communi, qua dicitur quod
sit actus corporis, magis est multipliciter dicti quam generis. ... Sed ex parte illa ubi propriam
naturam et rationem intellectus investigat, ut ex *tertio eiusdem libri*, ibi plane vult ipsum non
communicare in materia cum aliis, et separari a corpore, et non habere organum, et
impassibilem, et esse potentiam sine materia. Et omnia praedicta ostendit non solum de

This would imply that, although Siger has some doubts, he is not yet convinced that, as the theologians claimed and as his anonymous colleague conceded, Aristotle's position, as he has presented it, is heretical. During the next two years he would change his mind about this.

In chapters 4 and 5, Siger seems to be interpreting Aristotle as teaching the uniqueness of the intellect. He concludes that since the intellective soul is without matter and cannot be corrupted, it is eternal both in the past and in the future. But even though it is eternal, it is nevertheless caused. And although it was made, it is true to say that it was not made from anything preexisting. It is not, however, true to say that it was made from nothing, according to any meaning of the preposition 'from' (*ex*). If 'from' is taken in a causative sense, it is not true because "nothing is the cause of nothing." It is not true in the temporal sense, because it is eternal. It is not true in the sense of duration or natural order, because if it were actually nothing by its nature before it became something, it could not have received being even from another. "And I understand the intellective soul to be an eternal being (*semper esse*) of itself in this way: because in its *ratio* and definition it is an eternal being, since it lacks matter. But it does not have this eternal being, which is of its *ratio*, from itself effectively, but from another."[59]

Chapter 6 is very carefully thought out and reveals Siger's doubts about whether the intellective soul is one or many. "What the Philosopher thought ... about the separation of the soul from the body," he begins, "is difficult. But it seems that he considered it not to be completely and totally separable from the body."[60] In the midst of his fourth argument he interjects a digression that well illustrates his ambivalence about Aristotle's teaching on the uniqueness of the intellect. "If someone should say that the Philosopher says that the intellective soul is separated from the other powers of the soul as the perpetual from the corruptible; and again, since it itself is perpetual and incorruptible, when the body of which it is the act is corrupted, it must

---

potentia, sed etiam de substantia animae intellectivae. Hoc dicimus sensisse Philosophum de unione animae intellectivae ad corpus; sententiam tamen sanctae fidei catholicae, si contraria huic sit sententiae Philosophi, praeferre volentes, sicut in aliis quibuscumque." Cap. 3, *ed. cit.*, p. 88.

[59] "Et intelligo animam intellectivam de se semper esse ens sic: quia in eius ratione seu definitione est semper esse, cum careat materia. Istud tamen semper esse, quod est de sui ratione, non habet ex se effective, sed ab alio." Cap. 5, *ed. cit.*, p. 94.

[60] "Quid senserit Philosophus ... circa separationem animae a corpore, difficile est. Videtur tamen quod senserit eam non esse penitus separabilem et totaliter ab omni corpore." Cap. 6, *ed. cit.*, p. 95.

necessarily be separated from that body and remain separate, it must be said, according to the explanation of the Commentator and perhaps the meaning of Aristotle,"[61] that although it would cease being the act of that body, it might well become the act of another, since the soul cannot exist without the body. Since we know, and Siger knew, that Aristotle dismissed the transmigration of souls as a 'Pythagorean fable,' he must in this place be considering the intellective soul to be unique, but with the qualification that this is the view of Averroes and *perhaps* of Aristotle. But in the next two arguments he seems to consider that each man has his own intellective soul: 5- every body begins to be; the soul is eternal; therefore if the intellective soul were totally separable from the body and itself began to be the act of the body, then for an infinite time in the past it would have existed as the act of no body; and 6- if the soul were separate from the body, since there have been infinitely many men in the past, there would now be an actually infinite number of separated souls, and this is contrary to what Aristotle said about the infinite in *Physics* 3, 5 (204a-b).[62] These arguments only make sense with the additional assumption that each man has his own soul.

Then once again there is a gratuitous digression, this time discussing the matter of rewards and punishments, but cautioning that "it is not our principal intention to inquire into the truth about the soul, but what the opinion of the Philosopher was concerning it."[63] The deeds of the soul, he says, according to Aristotle, are common to it and the body, and God (the Legislator) ought to bestow rewards and punishments on the whole composite by honoring those who do well and punishing malefactors, because otherwise there would be something evil and disordered in the universe. (Interestingly, Gregory the Great expressed much the same view.) But divine providence does not prohibit there being evil in the universe; and good deeds are their own reward, while evil deeds are their own punishment. He follows this with a reminder of the limits of philosophy, saying that those who are experts in certain matters have knowledge of them that the inexpert lack. "And therefore, although philosophers, who are not expert in the matter of souls totally separated, do not hold that they are separated, nevertheless those who are

---

[61] "Quod si aliquis dicat Philosophum dicere animam intellectivam separari ab aliis virtutibus animae sicut perpetuum a corruptibili, et iterum cum ipse sit perpetua et incorruptibilis, corpus autem cuius est actus corrumpatur, necesse est eam separari a corpore et manere separatam, dicendum est secundum expositionem Commentatoris et forte intentionem Aristotelis ...". Cap. 6, *ed. cit.*, pp. 97-98.

[62] Cap. 6, *ed. cit.*, pp. 98-99.

[63] "... nostra intentio principalis non est inquirere qualiter se habeat veritas de anima, sed quae fuit opinio Philosophi de ea." Cap. 6, *ed. cit.*, p. 99.

expert know of the separation of the soul and have revealed it to others. ... [This is a matter] to which the common reason of man does not rise, except by believing the testimony of the prophets."[64]

And finally, he notes that Aristotle does not list the intellectual soul among the eternal substances in *Metaphysics* 12, 8 (1073a-b), and he concludes from this that Aristotle did not consider that the intellect was ever separated from the body.

This is followed by the chapter on whether the intellective soul is unique or multiple, with which we began our discussion of this treatise, and two chapters in which he argues more fully his points of disagreement with Albert and Thomas. It is in his reply to Albert that he repeats, and to some extent amplifies, his account of the composite nature of the human soul. He holds that, according to Aristotle, the vegetative and sensitive souls are generated by being educed from the potency of matter, while the intellective soul is a divine thing and is from without; and therefore they cannot pertain to the same substance. These three powers "are not those of one simple form, but are a kind of composite of the intellective coming from without and a vegetative and sensitive substance educed from the potency of matter."[65] This multiplicity of different parts pertains to the *ratio* of the completed composite.

And since man is a natural composite more perfect than others, "it is not illogical or to be wondered at if he is less 'one' than other natural composites, which have only one simple form or perfection."[66] We have noted this same view earlier in the *Summa de bono* of Philip the Chancellor and the *Summa de anima* of John of La Rochelle .

The work ends on a tentative note with an admonition to his students to be alert, "so that you might be aroused by the doubt remaining in you to study and read, 'cum vivere sine litteris mors sit et vilis (*sic*) hominis sepultura'."[67]

---

[64] "Et ideo, licet philosophi non experti operum apparentium de animabus totaliter separatis eas sic separatas non ponant, qui tamen experti sunt praedictam animae separationem noverunt et aliis revelaverunt. ... ad quae communis ratio hominum non ascendit, nisi credendo testimonio prophetae." Cap. 6, *ed. cit.*, pp. 99-100

[65] "... hoc non est simplicis formae, sed quodammodo compositae ex intellectu de foris adveniente et una substantia vegetativi et sensitivi educta de potentia materiae." Cap. 8, *ed. cit.*, p. 110.

[66] "... non enim est inconveniens neque mirabile si minus sit unum quam alia composita naturalia, quae non habent nisi unam formam simplicem seu perfectionem." Cap. 8, *ed. cit.*, p. 110.

[67] "... ut ex hoc dubio tibi remanente exciteris ad studendum et legendum, cum vivere sine litteris mors sit et vilis hominis sepultura." Cap. 9, *ed. cit.*, p. 112.

SIGER, *IN DE CAUSIS*. In 1966, A. Dondaine and L. J. Bataillon discovered a remarkable and very revealing set of questions on the *Liber de causis* by Siger,[68] a work that appeared in an excellent critical edition by Antonio Marlasca in 1972.[69] According to its editor, it was composed between 1274 and 1276 and most likely was based on a course Siger had given in the Parisian arts faculty and only partially revised for publication. Because this was a set of questions on the text and not an exhaustive commentary, the author was able to choose those matters he wished to discuss. This is important to keep in mind, since in this work Siger devotes much space to investigating eternity and the rational soul, two of the principal subjects of dispute since 1266. They reveal a scholar who has thought hard about these problems, has seriously modified his earlier positions, has lost much of his earlier faith in philosophy, Averroes, and even Aristotle, and to whom it has finally occured both that on the matter of the uniqueness of the intellect Aristotle's text is not at all clear, and that the position of Averroes on this point is heretical; and not only is it heretical, it is also irrational.

But not all of Aristotle's or Averroes's dubious opinions could be dismissed as irrational. After concluding in question 12 that in Aristotle's thought it was absolutely necessary that the celestial bodies and motions be eternal and separated substances, he adds:

> But nevertheless, because the authority of the Christian faith is greater than all human reason, and even the authority of philosophers, let me say that an Intelligence is not in eternity, although I do not have a demonstration for this. And because an Intelligence is not in eternity, the arguments designed to prove that it is in eternity are not necessary. Therefore one should try to refute them in some way.[70]

He follows this with two attempts to refute them and concludes: "But one should piously, firmly, and without further investigation believe the

---

[68] A. Dondaine and L. J. Bataillon, "Le manuscrit Vindob. lat. 2330 et Siger de Brabant," *Archivum Fratrum Praedicatorum* 36 (1966), 153-215.

[69] Antonio Marlasca, ed., *Les quaestiones super librum De causis de Siger de Brabant* (Louvain/Paris, 1972). Philosophes Médiévaux 12.

[70] "Nihilominus tamen, <quia> auctoritas fidei christianae maior <est> omni ratione humana et etiam philosophorum auctoritate, dicamus intelligentiam non esse in aeternitate, licet ad hoc demonstrationem non habeamus. Et quia etiam ex quo intelligentia non est in aeternitate, rationes volentes concludere eam esse aeternam non sunt necessariae, ideo temptandum est eas aliqualiter dissolvere." *Quaestiones in De causis* 12, *ed. cit.*, p. 66.

explanation that rests on the authority of the Christian faith."[71] He had also decided that, although Aristotle's natural science was exclusively concerned with transmutation in a subject, it was possible to reconcile creation with his teaching: "For it is indeed possible, even according to Aristotle, that some things may begin to be without a becoming that is theirs essentially, and by a mutation that pertains to them essentially; rather they are always found either in potency or in act, and never in potency mixed with act, which the *ratio* of motion requires."[72] And creation is not a true transmutation: "For creation, although it is understood by us in terms of the *ratio* of a kind of transmutation, is nevertheless not a transmutation. For in those sensible things that are proportioned to our intellect, thus it is that things newly made have been brought about by a transmutation and therefore we understand created things to have been brought about according to this notion of being made; but it is not thus, since they have been brought about from nothing and without a subject."[73] This sensitivity to the difference between philosophical and theological language is also revealed in question 25, where he says: "That which we have said, that all things that have been brought about here below are reduced to the first cause and that there is nothing new either in the soul or in the will or in other things <brought about> immediately by the First Cause, must be understood according to the common and natural meaning of things being brought about, <and> not referring to the miracles and prodigies of the omnipotent God, caused immediately by God."[74]

Although Siger was undoubtedly led to reconsider his position as a result of Aquinas's *De unitate intellectus*, he did not become a convert to Thomism.

---

[71] "Rationi autem illi quae innititur auctoritate fidei christianae firmiter et pie sine ampliore inquisitione credendum est." *In De causis* 12, *ed. cit.*, p. 67.

[72] "Bene enim est possibile, etiam secundum Aristotelem, quod aliqua esse incipiant absque fieri quod sit eorum secundum se, et mutatione quae sit ad ea secundum se; immo semper inveniuntur vel in potentia vel in actu et numquam in potentia actui permixta, quod exigit ratio motus" *In De causis* 20, *ed. cit.*, p. 67.

[73] "Creatio enim etsi intelligatur a nobis sub ratione cuiusdam transmutationis, non tamen transmutatio est. In his enim sensibilibus quae sunt proportionata nostro intellectui, ita est quod nova facta per transmutationem facta sunt, et ideo creata intelligemus facta secundum illam rationem factionis; non tamen est ita, cum ex nihilo et sine subiecto facta sint." *In De causis* 20, *ed. cit.*, pp. 88.

[74] "Quod autem dicimus omnia quae fiunt hic inferius reduci ad causam primam et nihil esse novum nec in anima nec in voluntate nec in aliis a causa prima immediate, intelligendum est secundum communem sensum et naturale fieri factionis ipsarum rerum, non intendente miracula et prodigia Dei omnipotentis immediate a Deo causata." *In De causis* 25, *ed. cit.*, p. 102.

He agreed with him on the matter of the unity of substantial form, and in question 4 he says that although an individual may receive the predication of genus and species and many differentia, "this is not because in its substance it has various degrees of forms. Rather, Socrates, through one single form, is a substance, a body, a living body, and a man, but under one *ratio* and another."[75] It is this last phrase that he uses in question 18 to distinguish among the vegetative-sensitive soul and the intellective, although so far as I can see he never achieved complete clarity on this. He denies that something that is by its nature separated from matter can be of the same species as that which is not separated, since the conjunction with matter follows form not accidentally, but through its essence.

He goes into the matter more fully in question 26, "Utrum anima humana impressa sit corpori sicut forma et perfectio." Here he summarizes Aquinas's position from chapter 3 of *De unitate intellectus* but says:

> But this position cannot stand. For since the intellective soul is the form and perfection of man (as the truth of the matter is), it cannot be a separated potency in operation.
> For matter, which is a being through some form, is able to operate and does operate by the potency and operation of that form. Further, to posit understanding common to body and soul because the substance that is the first principle of its understanding is united to its matter and thus is in matter, is to say that the understanding of matter is a perfection. But this is not true, for understanding does not have an organ in which it is. ... Further, Aristotle in 1 *De anima* says how understanding is common to the soul and body of the entire composite, so that man himself understands ... because understanding does not occur without phantasms. This does not argue that understanding itself is common in the sense that understanding needs the body as a subject in which understanding might take place, but only as an object. ... But Aristotle does not hold that they are common in this way, but because the soul, in understanding, is not sufficient by itself but needs the body and the body's powers to which it is naturally united for the fulfilment of the proper kind of operation, since there is no understanding without phantasms; it is from this fact that understanding is common to body and soul. Whence, understanding is not common to the soul and body in the

---

[75] "... hoc non est quia in substantia sua habeat diversos gradus formarum. Immo Socrates per formam unam simplicem est substantia, corpus, animatum corpus et homo, sub alia tamen et alia ratione." *In De causis* 4, *ed. cit.*, p. 48.

same was sensing is. For sensing is common to soul and
body because its being is in matter and in an organ. But it
is not thus with understanding. ... The sense faculty is not
naturally united to the object from which it derives its sense
impressions; but understanding is naturally united to the
body and bodily powers from which it has understanding.
... But it must be understood that the intellective soul is the
perfection and form of the body, but not like a vegetative or
sensitive soul. For the intellective soul thus perfects the
body that it also subsists by itself in its own being, not
depending on matter, not educed from the potency of
matter.   But the vegetative and sensitive souls are
perfections of matter such that they do not subsist by
themselves, and they do depend on matter in their own
being, since they are educed from the potency of matter
through the generation of the composite, by the
transmutation of matter toward its act and perfection."[76]

Siger's major disagreements with Aquinas were over the way the soul was
united to the body, and in what sense it is true that man understands. Aquinas
had held that although the soul was united to matter in its essence as form or
perfection, that the soul's power, or intellect, is separate from matter and in no

---

[76] "Sed haec positio stare non potest. Cum enim intellectiva anima sit hominis forma
et perfectio, sicut rei veritas est, non potest esse potentia et operatio separata. Materia enim,
quae est ens per aliquam formam, potest operari et operatur potentia et operatione illius
formae. Praeterea. Ponere intelligere commune corpori et animae quia substantia quae
primum principium est illius intelligere sit unita materiae sicut cuius est, sic in materia, est
dicere quod intelligere materiae sit perfectio. Hoc autem non est verum: non enim habet
intelligere organum in quo sit. ... Praeterea. Aristoteles *primo De anima* dicit qualiter
intelligere sit commune animae et corpori totius coniuncti, ita quod homo ipse intelligat, ...
quia intelligere non est sine phantasmate; sed intelligere non esse sine phantasmate non arguit
ipsum intelligere esse commune ex hoc modo quo intelligere egeat corpore sicut subiecto in
quo sit intelligere, sed tantum sicut obiecto. ... Non sic autem ab Aristotele ponitur commune,
sed quia anima intelligendo non sibi sufficit per se sed eget corpore et viribus corporeis
quibus naturaliter est unita ad propriae speciei operationis expletionem, cum non sit intelligere
sine phantasmate: hinc est quod intelligere est commune. Unde intelligere non est commune
animae et corpori sicut sentire. Sentire enim sic est commune animae et corpori quod est in
materia ens et in organo; non sic autem intelligere; ... sensus enim non naturaliter unitus est
obiecto ex quo debet sentire, intellectus autem naturaliter est unitus corpori et viribus
corporeis ex quibus habet intelligere. ... Sed est attendendum quod anima intellectiva est
corporis perfectio et forma, non sic tamen sicut vegetativa et sensitiva. Anima enim
intellectiva sic corpus perficit quod et per se subsistit in suo esse non dependens a materia, de
potentia materiae non educta. Vegetativum autem et sensitivum sic sunt materiae
perfectiones quod per se non subsistunt et in suo esse dependent a materia, cum de potentia
materiae educantur per generationem compositi, per transmutationem materiae ad suum
actum et perfectionem." *In De causis* 26, *ed. cit.*, pp. 105-06.

way the form of the body. Siger's final position was that both the substance
and power of the soul are the act and perfection of matter. But it is difficult to
see how one can reconcile this with his continuing insistence that the intellect
is absolutely independent in its essence from matter.

Still, he tries. The soul does not depend on matter for its subsistence, but
its nature is to be the form and perfection of the body. The nature that is the
principle of intellectual operation is a form subsisting *per se*, so that not only
is it the very thing by which something is but is a being *per se* not needing a
material subject in its being, although it needs it in its operation.[77]

It is evident that Siger's attitude on the faith-reason relationship was from
the beginning different from that of Boethius. The latter had insisted on the
independence of philosophy to investigate questions that lay within its
competence. He defined this as nature, and the methods a philosopher used
were reason and experience. Boethius did not deny that God may be able to
do many things outside the realm of natural causation, but this sort of thing
lay outside the area of the natural philosopher. Siger, on the other hand, as
early as the pre-1269 *In tertium De anima*, made a careful distinction between
philosophy (which usually meant the works of Aristotle) and the truth of any
question. He distinguished between necessary demonstrations and probable
arguments and explained why Aristotle's conclusions followed from his
premises, but he never claimed that philosophical truth was to be preferred to
revelation. It was clearly his hope that the two could be made to agree, but
when they could not, one must accept the teaching of faith and hope to
improve his philosophizing.

---

[77] [N]atura quae est intellectualis operationis principium est forma per se subsistens,
ita quod non tantum ipsa est qua aliquid est, sed est per se ens non egens in esse suo materiali
subiecto quamquam etiam egeret in operatione." *In De causis* 52, *ed. cit.*, pp. 179-80.

CHAPTER NINE

THE EVE OF THE CONDEMNATION

The period from the late 1260s to 1277 is difficult to understand. The evidence is somewhat contradictory, and the inclination to tell coherent stories has often led modern historians to emphasize one or another aspect of that evidence. The most common scenario is that of a building of tensions, as a result of the Parisian 'Averroists,' from about 1266 to the 1277 condemnation of 219 propositions by bishop Tempier, signaling the victory of the 'traditionalists.'

Such a story can be well documented. We have already noted the shrill admonitions of William of Baglione and the somewhat more restrained strictures of John Pecham. Between 1267 and 1273, Bonaventure held three conferences, or series of sermons (*De decem praeceptis, De donis spiritus sancti*, and *Hexameron*) at Paris, which are masterpieces in their way but which also exacerbated doctrinal dissention, identifying the uniqueness of the intellect, the eternity of the world, and the animation of the heavens as the three most dangerous errors threatening the faith. And even the usually even-tempered Thomas Aquinas composed a testy treatise on the unity of the intellect against the 'Averroists.'

In 1270, bishop Tempier condemned thirteen propositions, including the assertion that there is one intellect for all men, that it is false to say that this man understands, and that the soul is corrupted with the body, and forbade any master to maintain them.[1] The reaction to this condemnation is puzzling. Apparently no one paid any attention to it; certainly Boethius of Dacia did not, nor did the anonymous master of Bazán, nor Siger (although Siger seems to have taken the warning to heart). The minister-general of the Dominican order, John of Vercelli, instead of enforcing the bishop's decree within the order, asked for advice from Robert Kilwardby and Thomas Aquinas as to what his policy should be, that is, whether the condemned propositions were in fact erroneous.[2] In 1272 dissention within the arts faculty itself is

---

[1] H. Denifle and E. Chatelain, ed., *Chartularium Universitatis Parisiensis* (2 vols., Paris, 1889), I, p. 499 (#441).

[2] The responses to these inquiries have been published by M.-D. Chenu, "Les réponses de saint Thomas et de Kilwardby à la consultation de Jean de Verceil, 1271," *Mélanges Mandonnet. Études d'histoire littéraire et doctrinale du moyen âge* (2. vols., Paris, 1930), I, pp. 191-222, portions of which are also printed in Chenu, "Aux origines de la 'science moderne,'" *Revue des sciences philosophiques et théologiques* 29 (1940), 210-17.

evidenced by that faculty's forbidding its members to consider certain questions that were the exclusive province of theology, and in borderline questions not to conclude differently than the theologians. Also in 1272 a young student of Thomas Aquinas, the first Augustinian master, Giles of Rome, published his *De erroribus philosophorum*.[3] And M.-T. d'Alverny has published a description of a mutilated manuscript, which reveals the intensity with which at least one person resented the encroachment of dangerous novelties into the Christian tradition.[4]

By 1276 someone had complained to the pope, who at that time was the former Aristotelian scholar, Peter of Spain, about these novelties, and the pope thereupon requested the bishop of Paris to investigate the charges and report back to him. In November of that same year, the Inquisitor of France, Simon du Val, summoned Siger of Brabant and two of his associates to appear before the tribunal, but Siger, rather than obeying the summons, fled for protection to the pope.

All of this is well established, but there is more to the story. In the first place, the battle over dangerous theological novelties was not the most important conflict convulsing the University of Paris during this period. The renewed outbreak of hostilities between the mendicants, anxious to preserve and justify their privileges, and the seculars, jealous of these privileges and hoping to end them, generated much more high feeling and a more voluminous literature.[5] In the second place, the self-styled defenders of tradition were themselves simply an alternative version of the novelties of which they complained. They adopted Aristotle as the 'Philosopher.' They derived the doctrines of plurality of forms and universal hylomorphism from Avicebron, even while attributing them to Augustine. And much of their thought was more indebted to Avicenna than to the tradition of the church. And in the third place, not all masters, of either arts or theology, were actively engaged in controversy. Walter of Bruges, Eustace of Arras, and William of Falegar were all men of sharp minds, good will, and conciliatory disposition, and they were all Franciscans.[6]

---

[3]   Giles of Rome, *Errores philosophorum*, ed. Joseph Koch, tr. John O. Riedl (Milwaukee, Wisc., 19-44).

[4]   M.-T. d'Alverny, "Un témoin muet des luttes doctrinales du XIIIe siècle," *AH-DLMA* 17 (1949), 223-48.

[5]   On this see Decima Douie, *The Conflict Between the Seculars and the Mendicants at the University of Paris in the Thirteenth Century* (New York, 1954).

[6]   On these three men, see R. C. Dales, *Medieval Discussions of the Eternity of the World*, pp. 117, 118-20, and 129-32.

MATTHEW OF AQUASPARTA. Another excellent example of this type of scholar was the illustrious Franciscan master, Matthew of Aquasparta, who disputed a series of thirteen questions on the soul[7] while he was regent master in theology at Paris from 1275 to 1277. They constitute a comprehensive review of the more important opinions on various problems involving the soul from Plato to his own day, a classification of these views, a presentation of the arguments they are based on, and an evaluation of each position, all presented with exemplary clarity and economy. This is a theological, not a philosophical, work, although Matthew was well-read in the philosophical literature and competent in philosophical techniques. His mind was not keenly analytical or original, like those of Siger, Boethius, or Aquinas, but rather comparative and adjudicating. His stance is middle of the road: he opposes Aquinas as well as monopsychism, but one would never deduce from these questions that there was a raging controversy over monopsychism at the time. Matthew displays considerable independence and rejects several of the views of Gregory of Nyssa and other Christian saints, as well as those of many moderns, including John Pecham, whom he would succeed in 1279 as lecturer at the papal curia; and he shared several views with Siger.

These questions have been carefully reworked for publication. Matthew's Responses are carefully wrought essays, treating in each case the entire range of opinion on each question. He sometimes comes to a clear conclusion of his own, but he also sometimes grants that several of the competing answers are accepable. In spite of this lack of dogmatism, one can infer a systematic, if ultimately incoherent, doctrine from his arguments.

Most noticable among Matthew's views is the emphasis he places upon human unity. The soul is united to the body not as a prisoner, or as a mover, but as its form and perfection, so that together they make one thing.[8] Like Grosseteste, he emphasizes the soul's strong desire to be united to the body and its unwillingness to be parted from it, since this separation is contrary to nature.[9] It is nevertheless a subsistent being and is able to exist temporarily

---

[7] *Matthieu d'Aquasparta, Quaestiones disputatae de anima XIII*, ed. A.-J. Gondras (Paris, 1961). All page numbers are to this edition.

[8] "... quod spiritus intellectualis unitur corpori, non sicut carceri, nec sicut motor tantum mobili, sed sicut forma materiae et sicut perfectio perfectibili, ita quod ex eis fit vere unum, non tantum in agendo, sed etiam in essendo, quod ergo dicitur homo vere ex anima et corpore subsistens." Q. 2, Resp. II, p. 30.

[9] "Et primo, ex ipsa sui naturali inclinatione et appetitu ad corpus quod invite deserit. Et hoc manifestatur in ipsa separatione a corpore. Quod enim innaturaliter sive contra naturam ab altero separatur, naturaliter sibi unitum est. Unio autem naturalis corporis cum spiritu intellectuali non potest esse nisi sicut perfectibilis cum sua perfectione. Et indi est, ut

in separation from the body. Matthew investigates its status as a *hoc aliquid* in a manner reminiscent of, and probably indebted to, William of Baglione, saying that it fulfills three of the four necessary conditions of a *hoc aliquid* but lacks the fourth, independent being, because its natural inclination is to be the form of the body, and while it is not, its being is incomplete and its natural desire is frustrated.[10] And, again like Grosseteste, he considers man to be a microcosm summing up the entire universe, from lowest to highest, in his own being, "so that there might be made a certain most beautiful circle, joining the lowest and the highest."[11]

As the foregoing would indicate, Matthew upholds the necessity of positing a plurality of forms in man. He is emphatic and explicit on this, and it is in his notions of matter and form that he is unalterably opposed to Aquinas. He equates Aristotle's 'active potency' (*Metaph.* 7, 1034a and 9, 1046a) with Augustine's 'seminal reasons' (*De Gen. ad litt.* 9, 17),[12] and he accepts Bonaventure's teaching on the seminal reasons. Matter, he says, must be predisposed by earlier forms (of the elements, mixtures, complexions, etc.) to receive higher forms, by which it is ennobled. Even the soul contains matter, but this matter is ennobled to so high a degree that this does not constitute a hindrance to the soul's behaving as the form of the body. In fact, the whole distinction between matter and form seems to get lost in some of Matthew's arguments. The soul, however, is not the form of prime matter, which is only suited to receive first form, but of the body that has been previously organized by lower forms so that it is capable of being the 'matter' of the soul's 'form.'

---

dicit Augustinus, et supra allegatum est, quod anima separata non habet perfectum naturae suae modum, et quod inest sibi appetitus corporis administrandi, quo retardatur ne feratur totaliter in ipsum summum caelum." Q. 2, Resp., II, B, 1, p. 32. See also Q. 6, Resp., II, 3, p. 101.

[10] Fr. Brady has also called attention to Matthew's use of William's questions in other connections. See "The Background," pp 7, 35 (on Q. 7), and 42 (on Q. 7).

[11] "Cum autem in universo sit natura corporalis quae est infima, et natura intellectualis quae est suprema, oportet ponere aliqua naturam mediam in qua utraque consistat, ita quod fiat quidam circulus pulcherrimus dum infimo coniungitur ultimum. Hoc autem fit in homine in quo utraque natura convenit." Q. 2, Resp., II, D, 2, p. 35. See R. C. Dales, "A Medieval View of Human Dignity," pp. 569-71 for Grosseteste's expression.

[12] "Istud autem ens in potentia non est materia nuda, quia ex essentia materiae forma non fit; ergo si sit forma in potentia, quae cum non sit a datore, oportet quod sit in materia; et istud vocat Philosophus potentias activas, sed Augustinus vocat rationes seminales; oportet ergo illas rationes seminales sive illas potentias activas praeexsistere in materia, quae excitantur per agens extrinsecum, ut sint in actu; et ista sunt fundamenta formarum corporalium materialiumque, et corrumpi non possunt; ergo nec abici, advenientibus formis ulterioribus, ut ipsi ponunt." Q. 4, Resp., II, D, p. 67.

And Matthew does not attribute this position to Avicebron, but to Aristotle, and especially Averroes, who says that "all things between first matter and last form are composed of matter and form."[13]   So the soul, as many earlier masters had held, is not the form of the body insofar as it is a body, but insofar as it is the living composite. According to Aristotle's definition, the soul is the perfection of an organic body, and prime matter is not an organic body.

But it is the rational soul that places man in his specific being.   The preceding forms, though necessary, are incomplete, and the matter they inform is still in a state of flux until the arrival of the last form, the rational soul, which constitutes man as man. Matthew takes explicit exception on this point to Averroes, who makes the imagination or formal intellect the first perfection of man and holds that man is corruptible not only insofar as he is an animal, but also insofar as he is a man.   It follows from Averroes's position, Matthew says, that "man is not a rational animal, except perhaps *per accidens*, which is against all philosophy and reason ... and that man is not in the image of God and not capable of beatitude."[14]   And he is nearly as hard on Hugh of St. Victor (referred to only as a "modern philosopher") for saying that although man would not be man except through the rational soul, he is constituted in the being of an animal by the sensitive soul and his vegetative specific being by the vegetative soul.   "This opinion," says Matthew, although it is not so erroneous as the preceding and indeed contains something of truth and something of falsehood, has nevertheless more falsehood than truth."[15]

It is this point (i.e., the relationship of the vegetative, sensitive, and rational souls in man), which is the major weakness of any explanation depending on the plurality of forms and duality of substance, that Matthew spends more time on than on any other, and question 6 is entirely devoted to: "Whether in man the vegetative, sensitive and intellectual souls are of the same substance, or are different, so that the vegetative and sensitive are

---

[13]   "Nam anima non est actus primae materiae, sed est 'actus corporis organici,' ut dicit Philosophus, et cum hoc concordat Averroes, super I *Physicorum*, in principio, ubi dicit quod 'omnia inter primam materiam et ultimam formam sunt composita ex materia et forma'." Q. 4, Resp., II, F, p. 67.

[14]   "Sequitur enim ad hanc positionem quod homo non sit animal rationale nisi forte per accidens, quod est contra omnem philosophiam et contra rationem. [Nascitur et ulterius maius inconveniens,] quod homo non sit ad imaginem Dei, et quod non sit beatificabilis." Q. 5, Resp., I, B, p. 82.

[15]   "Ista autem opinio, licet non sit erronea sicut praecedens, immo contineat aliquid veritatis et probabilitatis, habet tamen magnam partem falsitatis et improbabilitatis admixtam." Q. 5, Resp., I, C, p. 83.

educed from the potency of matter, and the intellectual is created immediately by God." Matthew outlines five major positions on the question and evaluates each of them. The first, that there are three souls in man differing not only in substance but also in location, he quickly dismisses. But the second (which is essentially that of Grosseteste but probably refers specifically to Anonymous Van Steenberghen) is more interesting to him: that the whole soul is created all at once and infused into the body at conception; it has all the powers of the soul, some of which are material, some less material, and some completely immaterial, but exercizes them consecutively, not because of any defect on its part, but because of the ineptitude of matter. "This position," he says, "although superficially it seems sufficiently beautiful and probable, is erroneous or close to erroneous, as is clear to anyone who inquires into it closely."[16] His first objection is that it is contrary to divine law, since in Exodus 21:22 (he says Leviticus) we are told that if a pregnant woman is stoned and dies, the matter is treated as a homicide; but if she lives and the fetus is killed, it is not. His second objection is that it is contrary to the Catholic faith, since the soul will be reunited to its body at the Resurrection, and, if the fetus should perish before it becomes truly a human body, therefore there would be some form without matter, which is false. In the third place, it is against philosophy, since the soul is the essential part of man, which, with the body, makes truly one thing. But a part never has complete being except in the whole. Therefore, on the same assumption as the previous objection, the soul will eternally lack its natural perfection, which is absurd. And finally, because the soul has a natural and essential inclination toward the body (for otherwise it would not be the body's form); and if this appetite will never be fulfilled, it will be in vain; which is false and against the Philosopher.[17]

He says furthermore that it is ridiculous that the soul should be created and not be the act of any body, that the intellect should exist and not perfect man, and that a form should await the preparation of matter, since matter is rather disposed according to the requirements of form. But in this case, the creation of the intellective soul would depend on the body, whose act and perfection it is. Therefore it seems absurd that the soul should be created before it would be the perfection of the body; it neither is nor can be the perfection of the body except in an already formed body, which is, according to the teaching of Augustine (*Liber 83 quaestionum* 56), on the forty-fifth day.

---

[16] "Sed ista positio, licet superficietenus videatur satis pulchra et probabilis, tamen intrinsecus consideranti apparet manifeste quod est erronea vel errori propinqua." Q. 6, Resp., II, 2, p. 101.

[17] Q. 6, Resp., II, 3, pp. 101-02

The third position is that of Aquinas (not named) that there is one substance of the soul in one man, and that is the intellective, which is created immediately by God and performs all the functions that the vegetative and sensitive do in beasts, when the body has been organized and formed. "Although this position has something of truth," says Matthew, "nevertheless the manner of positing it is completely improbable, ... because it is necessary to posit in one and the same thing a plurality of forms, not through aggregation and accumulation, but through a certain complexion and order. Otherwise the last form would give complete being, or it would be the perfection of prime matter, and the intellective soul would give to the body its being as a body or corporeal being; which is completely absurd and, so I think, has already been sufficiently disproved above."[18]   And to the contention that the vegetative soul is corrupted when the sensitive soul arrives, and the sensitive when the intellectual arrives, Matthew says, as did Siger, Anonymous Admontensis and Anonymous Vaticanus, that this is manifestly false, because a thing is corrupted only by its contrary, and the higher forms are not the contraries of the lower, since the latter prepare and arrange the body for the former.[19]

The fourth position, which Matthew says is based on *De animalibus* 17 (i.e., *De generatione animalium* 2, 3, 736a), is that there are several material forms, not aggregated or accumulated, but rather ordered to each other, and there are several spiritual forms ordered so that each preceding form gives being to corporeal matter and perfects it to some degree, though not in an absolute sense (*simpliciter*). But the last form gives being *simpliciter* and makes it a *hoc aliquid*, because matter, up until the arrival of the last form, was in a state of flux and did not possess complete being. Then, adopting a position very similar to that of Philip the Chancellor and Siger, Matthew says that these forms, although they are in act with respect to matter, are nevertheless in potency with respect to further forms and are like material dispositions, behaving as matter. These are not three souls, nor are they three substances, because substance bespeaks a being complete all at once and *per se*. But they are like three substantial forms ordered as they are more or less perfect, but it is always the last (i.e., the intellectual) that gives complete being and makes it a *hoc aliquid*. The vegetative and sensitive are educed from the potency of matter and come from the parents, but the rational or intellectual is created

---

[18]   Q. 6, Resp., III, 4, p. 104: "Sed licet ista positio forte in se habeat veritatem, tamen modus ponendi est omnino improbabilis. Nam ... necesse est ponere in uno et eodem pluralitatem formarum, non per aggregationem seu accumulationem, sed per quandam complexionem et ordinem; alias forma ultima daret esse perfectum, sive esset perfectio primae materiae, et anima intellectualis daret corpori esse corpus sive esse corporeum, quod est omnino absurdum et supra, ut puto, iam sufficienter improbatum est."

[19]   *Loc.cit.*

immediately by God. Matthew judges this position, which he says is common to those who write according to philosophy, as "very probable and very rational," but he cautions that it seems to be contrary to the words of Augustine, who speaks not of three forms differing substantially, but of three powers. "It is not safe," he says, "for a professor of theology to hold a position contrary to the teaching of the renowned Augustine."[20] This would seem to be a more egregious assertion of the doctrine of the double truth than one can find in any of the 'Averroists.'

The final position, he says, is common to the theologians: "that the soul is one in substance, having these three powers consubstantial and connatural to itself, through which it gives life, sense, and intellectual being, so that the soul, according to its entirety and insofar as it is created by God immediately with all its powers and is immediately infused into a perfectly organized and formed body. And this is the intellectual soul," which is one soul that performs the vegetative, sensitive, and intellectual functions.[21]

But this leaves unanswered the important question of the relationship between the soul and whatever powers prepared the embryo for the soul's infusion. Matthew lists a number of alternative explanations without declaring his preference, and he concludes that the first three positions he has outlined (including that of Aquinas) are to be rejected, but the prudent investigator may choose whichever of the last two pleases him more, "provided that he does not exceed the limits of faith."[22] And so, after this long and exhaustive investigation of a crucial question concerning the soul's relationship to the body, Matthew leaves us up in the air. He insists that the whole soul, only partly operational and being partly material, cannot be infused into the embryo at conception. The body must be previously organized before the soul can become its form. He cautions that it is not safe for a professor of theology to oppose Augustine's view that the soul is one

---

[20] "Quamvis autem ista positio sit multum probabilis et communis videatur esse omnium philosophantium secundum veram philosophiam, tamen quia videtur esse contraria dictis Sanctorum et maxime Augustini, qui videtur dicere contrarium, ubicumque loquitur de ista materia, loquitur non tamquam de tribus formis substantialiter differentibus, sed de tribus potentiis, ideo non est tutum professoribus theologiae positionem hanc contrariam dictis doctionibus egregii Augustini tenere." Q. 6, Resp., IV, 4, pp. 107-08.

[21] "... quod una est anima in substantia, habens istas tres potentias sibi consubstantiales et connaturales, per quas vivificat, et sensificat, et dat intellectuale esse, ita quod anima secundum se totam et quantum ad omnes sui potentias immediate a Deo creatur et creando infunditur corpori formato et organizato perfecte; et ista est anima intellectiva." Q. 6, Resp., V, p. 108.

[22] "Primis enim tribus opinionibus reiectis, eligat inter istas duas prudens investigator quae magis sibi placuerit, dum fidei limites non excedat." Q. 6, Resp., V, 4, p. 111.

substance with three powers, but in this case it is not at all clear how the body is prepared for the infusion of the soul, what the nature of the preceding forms is, and how they are related to the soul. And his own preferred philosophical position is that of a plurality of forms in each substance. So Matthew's view is ultimately incoherent.

Question 7 asks whether there is one intellectual soul for all men. Matthew begins by pointing out that, according to Averroes, there are four differentiae of the intellect: the passive, i.e., the operation of the phantasms and imagination, which perishes with the body; the possible, or natural, intellect, which is able to become all things; the agent, which brings about understanding in act by abstracting species from material conditions; and the formal intellect, also called the speculative or intellect *in habitu*. He first discusses the various opinions concerning the agent intellect. Some say that it is the divine light that is the principle of knowledge. This, he says, is partly true and partly false. It is true that there is such a divine light, but it is false that there is not some power of the soul that abstracts and illuminates species.

Others, particularly Avicenna, say that it is some pure created Intelligence or angel, last in the order of Intelligences, which impresses intelligible species on individual minds. "This position," says Matthew, is Catholic according to one way of understanding it, but according to another way it is wholly erroneous ... for only God is superior to the rational soul, which is created in his image.[23] (This last position seems indebted to Grosseteste.) And third is the opinion of Averroes, that the agent intellect is unique for the human species. He puts off a discussion of this position to his remarks on the possible intellect.

Among the positions on the possible intellect, he first notes that of Alexander of Aphrodisias (via Averroes) that each man has his own, but it perishes with the body. Then he discusses "the error of Averroes, who says that not only the agent intellect, but also the possible, is one for all men."

Averroes, he says, "supposes that man is corruptible not only insofar as he is an animal, but also insofar as he is a man. But if man is generable and corruptible insofar as he is a man, it must be that the first perfection of man is generable and corruptible. Because he supposes that, according to Aristotle, the agent and possible intellects are ungenerated and incorruptible and completely immaterial and eternal, it necessarily follows that neither the agent nor possible intellect is the first perfection of man, and therefore it is not

---

[23] "Ista autem positio Avicennae secundum unum modum catholicum est, secundum alium modum est omnino erronea. ... [A]nima rationali, hoc ipso quod ad imaginem Dei facta est, solus Deus est maior sive superior." Q. 7, Resp., III, A, b, pp. 125-26. Cf. Grosseteste, *Hexaëmeron*, III, xiv, 7, *ed. cit.*, p. 151.

numbered according to the number of men."[24]   This position, says Matthew, is completely absurd and manifestly destroys the foundations of faith and Holy Scripture, because it destroys the creation (*conditio*), the Fall, and retribution. It also destroys the foundations of philosophy in that it posits the being of all men to be one. For if being is from form, therefore if there is one form for all men, there must be one being for all men, or, if it is distinguishable, it is only *per accidens*. This is one of the few instances in which Matthew misrepresents the position of an adversary, for Averroes did not hold that the intellect was the substantial form of man. Matthew does little to strengthen his case in repeating the argument that, on Averroes's view, all men would simultaneously know the same things. Albert the Great had accepted Averroes's explanation of this, and the refutation of Matthew's objection was by now a commonplace.

And after having taken us through the entire range of thought on the soul, Matthew presents a disappointingly banal statement of the Catholic position: "that both the agent intellect and possible intellect are numbered according to the number of individual men, because they are both the first perfection through which man is man; and although it begins with the body, so that the intellective soul is infused into a completely organized body as soon as it is created, nevertheless, because it does not depend on the body, therefore when the body is corrupted, [the soul] is separated and remains to live again."[25]

Although Matthew opposed monopsychism, he spent much less time and effort in refuting it than he did on many other aspects of the soul, and indeed he borrowed much of his treatment from William of Baglione. This question was clearly, for Matthew at least, though important, not central.

---

[24]   "Alia est positio vel, ut melius dicamus et verius, error Averrois qui ponit, sicut apparet manifeste, super III *De anima*, quod non tantum intellectus agens sed etiam possibilis est unus in omnibus hominibus, non numeratus secundum pluralitatem individuorum hominum.

Cuius positionis fundamentum est istud: supponit enim quod homo est corruptibilis non tantum secundum quod animal sed etiam secundum quod homo; sed si homo est corruptibilis et generabilis secundum quod homo, oportet quod prima hominis perfectio sit generabilis et corruptibilis; quia supponit, secundum Aristotelem, quod intellectus agens et possibilis est ingenerabilis et incorruptibilis et omnino immaterialis et aeternus, sequitur necessario ex hoc quod neque agens neque possibilis sit prima perfectio hominis, ideo nec numeratur secundum numerationem individuorum hominum." Q. 7, Resp., IV, B, pp. 127-28. Fr. Brady has pointed out the similarity that Matthew's treatment has to that of William of Baglione. See "The Background," p. 42.

[25]   "... quod tam intellectus agens quam possibilis numeratur secundum numerationem individuorum hominum, quia uterque est perfectio hominis prima per quam homo est homo, et quamvis cum corpore incipiat ita quod anima intellectiva creando infunditur et infundendo creatur in corpore perfecte organizato, tamen quia a corpore non dependet, ideo, corrupto corpore, separatur et manet ad iterum vivendum." Q. 7, Resp., V, p. 130.

Matthew owes a number of intellectual debts to an interesting variety of his predecessors. His debt to Grosseteste is clear not only by congruence of doctrine, but by means of expression. It was from the bishop of Lincoln that he derived his insistence on the soul's strong natural desire to be united to the body; his statement that in man all parts of the universe, from lowest to highest, are joined "in a certain most beautiful circle;" and his insistence that the human soul, created in the divine image, was the highest of all creatures. He filled out the first of these points by William of Baglione's presentation of the extent to which the soul was a *hoc aliquid*. From William of Auvergne he took the point that the soul does not give the body being as a body but may be considered the form and perfection of an already organized body. And it is interesting to note that he shared several characteristic positions with Siger of Brabant: that forms need not be simple, but one form may behave as matter in relation to a higher form; and that the lower, vegetative and sensitive, forms that organize the body cannot be corrupted by the arrival of the intellectual soul, since a thing is corrupted only by its contrary. And it is of considerable interest to note that Matthew was aware of the inconsistency between the docrine of plurality of forms and that of the unity of the soul, saying that the philosophers reason well and truly concerning the former, but that a theologian must not maintain any position that is contrary to Augustine; thus implying, it seems to me, the doctrine of the double truth. In fact, Matthew's most basic disagreement was with Aquinas, and it concerned Aquinas's insistence on the unity of substantial form. Some of his criticisms of Aquinas are quite cogent: for example, the points, which he shared with Siger, that a thing can be corrupted only by its contrary, and his denial that an immaterial form (the rational soul) could function also as a material form and give the body being as a body. But on other matters, the two masters were arguing from vastly different metaphysical assumptions, to some extent masked by a common Aristotelian vocabulary, particularly the meanings of the terms form, matter, substance, and substantial form. An argument becomes futile when the disputants mean different things by the common terms they use.

CHAPTER TEN

CONCLUSION

THE CONDEMNATION OF 1277. On March 7, 1277 bishop Tempier condemned 219 propositions, ranging over a large number of topics, imposing severe penalties on anyone who maintained them or even heard them.[1] Many of these were directed against the alleged 'Averroists' in the faculty of arts; a number seem clearly to have been directed against Aquinas;[2] and others against masters who have not yet been identified.[3] Much has been written about this condemnation, but there is still much we do not understand. It appears however that there was a good deal of misrepresentation of various masters' views, either through malice or stupidity.[4] Although the preface to

---

[1] "Ne igitur incauta locutio simplices pertrahat in errorem, nos tam doctorum sacre scripture, quam aliorum prudentium virorum communicato consilio districte talia et similia fieri prohibemus, et ea totaliter condempnamus, excommunicantes omnes illos, qui errores vel aliquem ex illis dogmatizaverint, aut deffendere seu sustinere presumpserint quoquomodo, necnon et auditores, nisi infra vij dies nobis vel cancellario Parisiensi duxerint revelandum, nichilominus processuri contra eos pro qualitate culpe ad penas alias, prout jus dictaverit, infligendas," *Chartularium Universitatis Parisiensis*, ed. H. Denifle and E. Chatelain I (Paris, 1889), # 473, p. 543. I use the numbering of this edition in what follows.

[2] See the doctoral dissertation of Henry F. Nardone, *St. Thomas Aquinas and the Condemnation of 1277* (Washington, D.C., 1963). The Catholic University of America Philosophical Studies No. 209, especially pp. 102-15.

[3] The most thorough study of the condemnation, which identifies the source of many of the condemned propositions, is Roland Hissette, *Enquête sur les 219 articles condamnés à Paris le 7 mars 1277* (Louvain/Paris, 1977). Philosophes Médiévaux 22.

[4] In a personal communication to me, Prof. Joseph Goering has suggested that we make a mistake in considering "the condemnations to have been aimed at the actual teaching of real masters. My reading of the condemnation (as a canonico/theological document) is that it is conciously prescinding from naming names and, as far as possible, from implying that anyone is actually teaching these things. The presupposition is that no one would do such a thing, and if they did, they would do it accidentally and unwittingly. It is like a warning, not like an (academic) polemic against colleagues. We tend to read it as a theological disputation rather than as a disciplinary document establishing authoritative boundaries within which the masters are free to manoeuvre at will. Thus one should be surprised to find in the 'condemnations' the actual words or teachings of a Master. And one should not think that the bishops have 'twisted' or misrepresented the teachings of specific masters—they are only saying what would be false should anyone actually say it." I am not convinced that he is right, but I think that this point of view deserves serious consideration.

the list of condemned propositions asserts that certain masters were teaching a doctrine of the double truth, and a marginal note in one manuscript identifies these masters as Boethius of Dacia and Siger of Brabant, no one has ever discovered a text that teaches such a doctrine, and it is the consensus of modern scholarship that no one ever did.[5] It has also been established that no one ever taught the beginninglessness of the world as truth.[6]

But what of the doctrine of the uniqueness of the possible intellect? This is a much more complicated matter, and some texts do indeed seem to teach it. In fact a total of twenty-four condemned propositions concern the human soul, and an additional four (numbers 40, 120, 145, 154) are concerned with the nature, status, and competence of philosophy. No fewer than eight propositions concern the uniqueness and/or eternity of the intellect (numbers 27, 31, 32, 96, 109, 121, 125, 193). This is a quintessentially Averroistic view, it is clearly heretical, and it was maintained by several masters whose works we have. But the condemnation takes no account of the sense in which it was being maintained: as the truth *simpliciter*; as the proper understanding of Aristotle; as the necessary conclusion of philosophy restricted to the realm of nature; or as a more probable of two undemonstrable positions. As I read the texts, that it was the truth in an absolute sense was maintained by no one, although each of the other senses is maintained by one or more of the masters in arts.

Although Averroes was uppermost in the minds of those behind the condemnation, the relation of the condemnation to the tradition of Latin thought on the soul during the thirteenth century is quite complicated, and at least some doctrines of the majority of the masters we have investigated would be heretical by its terms. The concentration of the condemnation on combatting Averroes led its authors to insist that the soul is the substantial form of the body and to condemn the position that the soul-body relationship was essentially operational (numbers 7, 13, 14, 118, 119, 123). This was indeed Averroes's doctrine. He taught that the form that distinguishes one individual from another is the vegetative-sensitive soul, the seat of the speculative intellect; and that this soul is united operationally to the unique intellect, which is in potency to the phantasms provided by each sensitive soul. But the dualism of most of the pre-Thomist thinkers led them also to claim that the body, as body, had a form or soul distinct from the one created and infused by God. Indeed, this was the basis of both Kilwardby's and Pecham's objections to the doctrine of the unity of substantial form, since

---

[5] See Richard C. Dales, "The Origins of the Doctrine of the Double Truth," *Viator* 15 (1984), 169-79, and the bibliography cited there.

[6] Richard C. Dales, *Medieval Discussions of the Eternity of the World* (Leiden, 1990), pp. 140-55.

unless the body had a form of its own apart from the rational soul, the body of the Lord and of the saints would not be numerically the same substance after death as in life. The view that the soul is the form of the body was shied away from by most of the masters we have investigated, and one, Alexander Nequam, denied it outright. Albert the Great had been uncomfortable with the expression, preferring to call the soul the body's perfection. The same is true of Bonaventure, although he did sometimes use the term, and in his *Sentences* commentary, book 2, d. 17, he provided an ingenious account of the way it could be true, even as soul and body were considered to be distinct substances. Those who adopted the word form for the soul always had to redefine form in order to maintain the separability of the soul. They would say that it is not only a form (*forma tantum*) but also a substance in its own right, and then try to explain how this was possible. This involved them in many difficulties and rendered virtually impossible any attempt to account for human unity. It would be difficult to imagine a more purely operational relationship between body and soul than that expressed by Alexander of Hales and William of Auvergne, or indeed Augustine's definition of man as a soul using a body. There was a tension between the desire to have the soul united substantially to the body and the conviction that soul and body were two distinct substances.

The dualism of body and soul was maintained by all masters prior to Aquinas, and except for Roland of Cremona, Peter of Spain, and Albert the Great, required that the body, as body, had a form of its own. Whether this was identified with the sensitive soul (Fishacre, William of Auvergne) or as the corporeal form (Alexander of Hales), it comes perilously close to the position of Averroes, and even more so when a double vegetative-sensitive soul is admitted, as it was by Alexander of Hales, Richard Rufus, and Bonaventure.

Although Averroes's position on the relation of the vegetative-sensitive soul to the unique intellect is not exactly analogous to the view that the human soul is compounded of corruptible and incorruptible constituents, the two views have much in common and present similar difficulties: if only a part of the human soul is immortal, how can the whole man (or the whole soul) be considered immortal? Since the 'traditionalists' paid lip service to Augustine's (and pseudo-Augustine's) doctrine of the unity of the soul but also adopted the theory of plurality of forms, while at the same time insisting that the rational soul was the form of the body, they put themselves in an untenable position, which seems to have been realized only by Matthew of Aquasparta.

There was then in the thirteenth century a range of solutions to the problem of the relationship between body and soul, each of which had its own difficulties. The possibilities are: 1- a double vegetative-sensitive soul, of which the prior remains as a medium of union after the infusion of the rational

(Alexander of Hales, Richard Rufus, Bonaventure); 2- the denial that the embryo has a life and soul of its own but lives by virtue of the soul of one or both of the parents, like an apple on a tree, in Alexander of Hales's simile (Alexander of Hales, Roland of Cremona, Peter of Spain); 3- the absorption of the lower forms by the higher, the lower ceasing to exist (William of Auvergne); 4- a succession of forms, each one either preparing the body for the arrival of the next or acting as matter in the new composite soul, and either remaining functional in the new composite (Philip the Chancellor, Roger Bacon, Anonymous Admontensis); or 5- a succession in which the lower form is corrupted by the arrival of the next higher one, which has everything the lesser had plus something more, so that there is only one substantial form at a time and it is simple, though possessing a number of powers (Aquinas, Albert the Great); and finally 6- the infusion of the whole soul at conception, performing successively the requisite functions (Grosseteste, Anonymous Van Steenberghen). Theories of a compound soul account for the development of the embryo into a human body before the infusion of the rational soul but have difficulty explaining how a soul partly corruptible and partly incorruptible can be truly one soul, and they ignore the possible requirements of the resurrected body. The Augustinian position of the unity of the soul avoids this problem but has trouble accounting for the development of the embryo. The position that the soul was a form of matter without being a material form did not adequately explain how it could have a power not involved in matter. And the view that the whole soul was infused at conception did not account for the consequences of the possibility that the fetus might perish before it reached the status of a human body.

THE 'LATIN AVERROISTS.' It has become increasingly clear during the past generation that there was not an 'Averroistic' party at Paris in the sense that there was a group of masters who followed Averroes blindly on all matters and who shared a set of doctrines that constituted the position of such a party.

There certainly were some who valued philosophy highly and claimed an independent sphere of inquiry for it, but we have no evidence that anyone claimed the possibility of a double truth, despite the allegation of the epistolatory preface to the 1277 condemnation.

I should like to reiterate here the distinction between the task of a theologian and that of an artist. The subject matter of the theological faculty dealt with the data of revelation, and the masters of that faculty used philosophy to rationalize this body of data, to draw out the implications of its content, and to settle questions concerning the faith whose answers were not immediately apparent. Their responsibility was great, for the establishment of the true faith, upon which depended salvation, was their goal. And if some of the questions they debated seem frivolous, many—indeed most—were not. The primary task of the artist, on the other hand, was to teach the doctrine of the book upon which he was lecturing. In his disputations, he may or may not be arguing a question that had anything to do with faith; but when it did, he was required to investigate it on the basis of human reason and experience.

While it would certainly not have been permissible for anyone to teach as the truth *simpliciter* a doctrine that was clearly heretical, there is no incontrovertible evidence that anyone did this. Natural philosophy was restricted to investigating the realm of nature. The study of the soul lay within the province of both the artists and theologians (and John Blund had claimed it exclusively for the artists), but, as all commentaries and sets of questions on the soul from the arts faculty point out, the artist could study it only insofar as we have experiential knowledge of it—that is, how it operates in conjunction with the body. (As Boethius of Dacia pointed out, this is the only way we know that the soul exists.) He could not consider it in separation from the body; and in fact philosophy could not show that it ever did exist apart from the body. It is this difference in the responsibilities of the two faculties that accounts for much of the misunderstanding between them.

In spite of this, there are times when a master of one faculty showed a tendency to cross over into the other's field. When the artists did this, it was sometimes a case of blatant trespass—for example, when they asked whether the separated soul could suffer from hell-fire, since we can have no experience of the soul in its separated state; and yet this question frequently appears in the works of the artists. More often the faculties treated matters that lay within the province of each, as indeed was the case with the soul. When theologians crossed over, it was usually to show that the conclusions of true philosophy were the same as those of faith. Such theologians as William of Auvergne, Albert the Great, and Thomas Aquinas explicitly stated in certain works that they were proceeding *per viam philosophiae*.

Just as there was a range of views concerning the relation of philosophy and theology (or reason and faith) among the theologians, so there was among the philosophers of the arts faculty. Among the theologians views ranged from the position of Grosseteste and Bonaventure that philosophy had no independent sphere, through the view of Aquinas that reason and revelation cannot be in conflict, but there are matters (such as the Trinity) to which reason cannot attain, to the more extreme position of William of Baglione and a few others that all articles of faith can be conclusively demonstrated. Among the philosophers there is the stance of Anonymous Bazán, who recognized that reason and faith seemed sometimes to reach different conclusions but made no attempt at reconciling them, although he implied that the arguments on either side were not demonstrative. An extreme position was that of Boethius of Dacia, who insisted that philosophy was competent to treat any question that might be investigated by human reason and experience, and that there cannot be a contradiction between its conclusions and those of faith, since the natural philosopher only investigated the order of nature, not things done by God's omnipotence outside the course of nature (and Aquinas concurred in this). And most revealing is the progressive development of the thought of Siger of Brabant, who correctly pointed out that his job was to ascertain correctly the meaning of Aristotle; who even as early as 1269 noted that Aristotle's views were not necessarily correct but were the necessary consequences of his assumptions; and whose mature position, as we find it in the questions on *De causis*, was that human reason is not a perfect instrument, that even the greatest philosophers make mistakes, and that when we conclude according to philosophy something contrary to faith, we must accept the truth of faith and try to improve our philosophizing, but without thereby violating the rules of philosophical procedure. As he says, a prophet can know things that are hidden from other men, and they are true, but philosophy and prophecy are different things.

Nor was there anything like unanimity among the 'traditionalists.' Although they all show the influence of Bonaventure to some degree, I have found no one who accepted his imaginative explanation of soul and body constituting one substance because of residual potency on the part of the body and residual act on the part of the soul. While most of them adopted a theory of plurality of forms, and therefore a compound soul, from the artists, the most extreme among them, William of Baglione, objected to any theory of a compound soul. Matthew of Aquasparta sharply criticized some of the positions of John Pecham and seems to have preferred the position of Anonymous Vaticanus to any other account of the soul. So while one may say that they all held to some sort of dualism, insisted that the soul was a *hoc aliquid* and was composed of spiritual matter and form, and was the substantial form of the body (although this last position placed them in opposition to the founder of their school, Alexander of Hales), one cannot say that there was a consistent Franciscan, or 'traditionalist,' doctrine of the soul.

There was in fact much continuity in the philosophical discussions of the soul, extending from John Blund and Philip the Chancellor, through Peter of Spain, Roger Bacon, Anonymous Admontensis, Anonymous Vaticanus, to Siger of Brabant and Boethius of Dacia. The major element of novelty in the tradition resulted from the realization during the 1250s that Averroes had understood Aristotle as having taught that the intellect was unique, and the adoption of this interpretation by some of the arts faculty as being at the least the correct understanding of Aristotle and at the most the necessary conclusion of philosophy investigating only the realm of nature.

TRADITION VS. NOVELTY. The portrayal of the disagreements from 1266 to 1277 as a conflict between tradition and novelty is largely derived from the preface to the condemnation itself, Giles of Rome's *De erroribus philosophorum*, and John Pecham's letters to the University of Oxford, November 10, 1284 and to the bishop of Lincoln, June 1, 1285, both written after he had become archbishop of Canterbury. In the first of these letters, written to forbid the maintenance of any of the propositions forbidden by Kilwardby until the matter could be reconsidered, Pecham asserts the heretical nature of the theory of the unity of substantial form, a position he insinuates originated with two people who ended their lives wretchedly south of the Alps, although they were not born there.[7] In this letter Pecham took pains to point out that the issue was not between Dominicans and Franciscans, since his predecessor Kilwardby was a Dominican. But in the second letter, written in reply to an attack on him by a Dominican preacher, after assuring the bishop that he is not opposed to the works of the philosophers insofar as they serve theology, he complains that during the preceding twenty years profane novelties, contrary to philosophical truth and hurtful to theology, had been introduced into theological teaching. In a long rhetorical question he contrasts the doctrine "of the sons of blessed Francis, such as brother Alexander of holy memory and brother Bonaventure and those similar to them," to those who teach the complete opposite of what Augustine teaches "about the eternal principles and the incommutable light, the powers of the soul, the seminal reasons embedded in matter, and innumerable similar things."[8]

This view has been aggressively challenged by two Dominican scholars, Daniel A. Callus and James A. Weisheipl,[9] and although their attack was

---

[7]   "... nec eam credimus a religiosis personis sed saecularibus quibusdam duxisse originem, cuius duo praecipui defensores vel forsitan inventores miserabiliter dicuntur conclusisse dies suos in partibus transalpinis, cum tamen non essent de illis partibus oriundi." *Registrum Epistolarum Fratris Johannis Peckham Archiepiscopi Cantuariensis*, ed. Charles Trice Martin, Rolls Series 77 (London, 1885, repr. 1965) 3, p. 842.

[8]   The whole passage is as follows: "Praeterea noverit ipse quod philosophorum studia minime reprobamus, quatenus misteriis theologicis famulantur; sed prophanas vocum novitates, quae contra philosophicam veritatem sunt in sanctorum injuriam citra viginti annos in altitudines theologicas introductae, abjectis et vilipensis sanctorum assertionibus evidenter.
      Quae sit ergo solidior et sanior doctrina, vel filiorum Beati Francisci, sanctae scilicet memoriae fratris Alexandri, ac fratris Bonaventurae et consimilium, qui in suis tractatibus ab omni calumnia alienis sanctis et philosophis innituntur; vel illa novella quasi tota contraria, quae quicquid docet Augustinus de regulis aeternis et luce incommutabili, de potentiis animae, de rationibus seminalibus inditis materiae et consimilibus innumeris, destruat pro viribus et enervat pugnas verborum inferens toti mundo?" *Ed. cit* 3, p. 901.

[9]   Daniel A. Callus, *The Condemnation of St. Thomas at Oxford* (Blackfriars [Oxford], 1946, second edition 1955); and James A. Weisheipl, "Albertus Magnus and Universal Hylomorphism: Avicebron. A Note on Thirteenth-Century Augustinianism," in *Albert the Great. Commemorative Essays*, ed. Francis J. Kovach and Robert W. Shahan (Norman, Oklahoma, 1980), pp. 239-60.

undoubtedly not disinterested, it is by and large accurate. The main points they make are that the matter-form composition of the soul and the theory of plurality of forms are not Augustinian but come from the *Fons vitae* of Avicebron; that the traditional understanding of the composition of spiritual substances was that they were composed of *quod est* and *quo est*, as taught by Boethius; that Augustine's position was unequivocally on the side of unity of form, so that the 'novelties' introduced into Latin thought were the positions held by the self-styled traditionalists, whereas Thomas and his followers were simply restating the traditional Christian position in a more precise and scientific fashion. They also strongly imply that Pecham is a liar in claiming that it was the "sons of blessed Francis" who were champions of what they considered true philosophical and theological teaching, since a number of Franciscans whose works Pecham used (Alexander of Hales, Richard Rufus, and John of La Rochelle) had maintained the unity thesis and eschewed the hylomorphic composition of souls; that he is incorrect in claiming Augustine as a teacher of the plurality of forms; and that he was a conscious calumniator in implying that the origin of the unity thesis lay in the doctrines of Siger of Brabant.   Fr. Weisheipl also emphasized the corruption of Augustine's teaching on divine illumination by Avicennist notions.

Although it seems to me that Fr. Callus pays too little attention to the duality of body and soul, which dominated most thought on the soul before Aquinas, and that he considers unity of the soul to be equivalent to unity of substantial form (since, as we have seen, many early scholastics would not accept that the soul was a form), still the main thrust of his argument and that of Fr. Weisheipl is correct in pointing out that the theories of plurality of forms and the hylomorphic composition of the soul were indeed novelties; that their attribution to Augustine cannot be maintained without considerable arbitrary glossing of the text of the bishop of Hippo; and that posing the story in terms of tradition versus novelty is a deliberate falsehood, for which John Pecham bears the major responsibility. As I have asserted earlier in this study, the conflict was not between tradition and novelty but between two sets of conflicting novelties. Which set of new authorities was more consonant with the older tradition is still an open question.

SIMILARITIES BETWEEN BONAVENTURE AND AQUINAS. In the theological faculty, the disagreement over the soul was derived from the teaching of Bonaventure on the one hand and Aquinas on the other, and historians of philosophy would undoubtedly find their positions unalterably opposed. But if one looks behind the technical vocabulary, the views of Aquinas and Bonaventure have much in common. Bonaventure insists that the soul must contain spiritual matter, because he did not see how it could be individual and a *hoc aliquid* otherwise. Aquinas, while admitting potentiality in the composition of the soul, refused to call this matter and attempted to solve the problem of how the soul, a spiritual substance, could be individuated, by saying that it was individual because it was created to be the form of an individual human being. Bonaventure's followers, though not Bonaventure himself, misrepresented Aquinas as having said that the soul was individuated *by* the body *(a corpore)* rather than *ex corpore*, or by virtue of the body. As to the soul's status as a *hoc aliquid*, apart from the use of the term, both masters agreed that the soul was capable of existing after its separation from the body; that this existence was true but incomplete; and that the full being of the soul was being the form of the human composite. Both masters also stressed the soul's strong desire to be united to the body and the fact that it was ennobled, rather than imprisoned, by this union.

The metaphysical positions of the two were far apart (Bonaventure teaching the plurality of forms, seminal reasons, the hylomorphic composition of the soul, and the duality of substances; Aquinas teaching the unity of form, denying that the soul contained matter, and considering man to be a single substance), but their views on what the soul was were not so far apart as the technical expression of their systems made them appear. For both the soul was individual for each human being: it was immortal and therefore capable of existing apart from the body; its fullest existence was in union with the body; and it was ennobled and completed by this union. It is to be regretted that no one seemed interested in stressing these similarities and reconciling the differences.

THE DIMINUTION OF DUALISM. Throughout the thirteenth century, we find a gradual softening of the dualism that characterized the writers early in the century. The first important departure was Peter of Spain's implied denial that the body had a form of its own, and his explicit assertion that the only substantial form of a human being was the rational soul (although he was not completely consistent in this). Under the impact of Aquinas's teaching, even those who maintained the vocabulary of dualism moved in the direction of unity. Much of this seems indebted directly or indirectly to Robert Grosseteste, who had assumed a duality of body and soul but had used the term substantial form (though in a most un-Aristotelian sense) for the soul and had discovered the unifying principle of a human being in person, a notion that he picked up from one of Augustine's *obiter dicta*. He also stated in an extreme form the soul's desire to be united to the body, and this had echoes during the second half of the century. William of Baglione's insistence that the soul was a *hoc aliquid* sounds as though he is reverting to an outright dualism, but as he explains the nature of the soul, it becomes evident that it is not strictly speaking a *hoc aliquid*, since it lacks independent being and only has its full being when it is acting as the form of the body. Bonaventure had preceded him in this and had given an account of the body-soul relationship in which neither is complete without the other. Albert the Great sometimes seems to be almost a Thomist in his remarks on the unity of substantial form, but he could never completely overcome his conviction that body and soul are distinct substances. Only Aquinas and Anonymous Van Steenberghen achieved an explanation that made man truly one thing.

ATTITUDES TOWARD ARISTOTLE. It is even more true concerning Aristotle's doctrine of the soul than concerning his teaching on the eternity of the world that there was almost no hostility toward him. Only William of Auvergne took issue with him (as he understood him through Avicennan spectacles), and only Robert Grosseteste ignored him. It is possible that Grosseteste did not know the *De anima* (though I consider this unlikely), but he was certainly well acquainted with the *De animalibus*. Everyone also was willing to accept Aristotle's definition of the soul and most features of his account of it, and they disagreed only on what he meant, particularly concerning the agent intellect. And concerning the possible intellect, it was not realized until the mid-1250s that Averroes considered Aristotle to have held that it was unique. Once this was realized, it was considered a perversion of Aristotle by all but several of the natural philosophers in the arts faculty. But in spite of their attitude toward him, Aristotle was indeed the source of the disintegration of the traditional view of the soul. By considering it a form he posed for the scholastics an insoluble problem, which even Aquinas could not work out in a completely satisfactory manner.

Related to this is the variety of attitudes one finds, among both artists and theologians, on whether the Fathers spoke with authority, or simply as men, when they were discussing natural philosophy. Geoffrey of Aspall and Thomas of York (neither of whom we have treated in this work), Richard Rufus, Adam of Buckfield, Roger Bacon, and Albert the Great all explicitly stated that when the Fathers were discussing natural philosophy they speak only as men and can be refuted by arguments like any philosopher. Anonymous Vaticanus explicitly says that the position of Augustine should not be adhered to. Siger implies the same view when he says that the arguments of neither Averroes nor Augustine are demonstrative on the matter of the eternity of the intellect, but one cannot hold both positions; if one is right, the other is wrong. Some, such as Blund and Peter of Spain, simply avoided mentioning the Saints. And Matthew of Aquasparta claimed inspiration, for Augustine at least, when he is discussing philosophical matters (and even when his position is inconsistent with the philosophical views Matthew preferred).

UNEXPECTED CONCLUSIONS. Many of the results of this study have been no surprise. *Augustinisme-avicennisant* has been a part of the canon of the history of medieval philosophy for nearly a century. The importance of Bonaventure and Aquinas as the exemplars of two approaches to the soul was also a foregone conclusion. And it is almost like beating a dead horse to criticize the condemnation of 1277, although it is still instructive to see some of the details of its shortcomings. But some other things were not expected. The complete disarray of the concept of the soul by about 1235 was much more extreme than I had expected. Also, the importance of two largely overlooked masters, Philip the Chancellor and William of Baglione, has become clear, and this study has confirmed the picture that emerged from the investigation of debates on the eternity of the world that two of the most important figures in the history of thirteenth-century thought were Philip the Chancellor and William of Baglione. Now that Nicolas Wicki's edition of the *Summa de bono* is available and much of William's writing has been edited or described by Fr. Brady, these men will loom larger in future works on thirteenth-century thought. There is also a surprising amount of similarity between the positions of the artists and the 'traditional' theologians. A majority of both groups adopted the theory of plurality of forms and a duality of substance in the soul. They also often designate the sensitive soul as the form of the body as body, although Averroes considered it the form of man, and one Franciscan theologian, Matthew of Aquasparta, undoubtedly without realizing it, clearly implied a doctrine of the double truth. The term 'substantial form' was also used by most of the theologians in a manner similar to that of Siger in *De anima intellectiva* and the questions on *De causis*, that is, in an operational and completive or perfective sense, rather than in its true Aristotelian meaning. And William of Baglione agreed with the Averroistic position that if the soul lacked matter it would cease to be individual after its separation from the body.

But the most unexpected result of this study is the fact that among the theologians no one but Robert Grosseteste and among the artists no one but Anonymous Van Steenberghen held the position that since 1869 has been the official position of the Roman Catholic church: that the whole soul is infused at conception. In fact, no one but Roland of Cremona and Matthew of Aquasparta even mentioned the subject. Matthew cited Augustine's *De 83 quaestionibus* to the effect that the soul is infused on the forty-fifth day, and he specifically claimed that it would be contrary to divine law, to the Christian faith, and to sound philosophy for it to have been infused at conception. For the other masters, even though no one explicitly mentions it, the implication of their various doctrines is that the soul is infused at the point that the fetus is viable. That it could have been infused at conception is impossible in the schemes of all who held with a theory of plurality of forms or a succession of forms, regardless of the details of such a progression.

SELECT BIBLIOGRAPHY

Abbreviations:

AFP = Archivum Fratrum Praedicatorum
AHDLMA = Archives d'histoire doctrinale et littéraire du moyen âge
BGPM = Beiträge zur Geschichte der Philosophie des Mittelalter, ed. Clemens Baeumker
    (Münster i.W.. 1891 ff.)
CCL = Corpus Christianorum, series latina (Turnhout, 1953 ff.)
CSEL = Corpus Scriptorum Ecclesiasticorum Latinorum (Vienna and other cities, 1886 ff.)
PL = Patrologiae cursus completus ... Series latina, ed. J. P. Migne (221 vols, Paris 1884-64)
RNP = Revue néoscholastique de philosophie
RSPT = Revue des sciences philosophiques et théologiques
RTAM = Recherches de théologie ancienne et médiévale

Editions of Sources

Albert the Great. *B. Alberti Magni ... Opera Omnia*, ed. S. C. A. Borgnet (38 vols., Paris:
    Vives, 1890-99.
Alexander Nequam. *Alexander Nequam, Speculum Speculationum*, ed. Rodney M. Thomson,
    Auctores Britan nici Medii Aevi 11 (London, 1988).
Alexander of Hales. *Doctoris Irrefragabilis Alexandri de Hales Ordinis Minorum Summa
    theologica* (4 vols., Quaracchi, 1924-48).
Anonymous Admontensis. *Ein Anonymer Aristoteleskommentar des XIII. Jahrhunderts.
    Quaestiones in Tres Libros De Anima (Admont, Stiftsbibliothek, cod. lat. 367)*, ed.
    Joachim Vennebusch (Paderborn, 1963).
Anonymous (Bazán). *Ignoti auctoris Quaestiones super Aristotelis librum De anima*, ed.
    Bernardo Bazán, in *Trois commentaires anonymes sur le traité de l'âme d'Aristote*
    Philosophes médiévaux 11 (Louvain, 1971), pp. 351-527.
Anonymous (Giele). "Un commentaire averroïste du traité de l'âme d'Aristote," ed.Maurice
    Giele, *Medievalia Philosophica Polonorum* 15 (1971), 1-168 and *Un commentaire
    averroïste sur les livres I et II du traité de l'âme*, ed. Maurice Giele in *Trois
    commentaires anonymes*, pp. 13-120.
Anonymous, *La 'Summa Duacensis' (Douai 434)*, ed. P. Glorieux (Paris, 1955).
Anonymous (Van Steenberghen). *Un commentaire semi-averroïste du traité de l'âme*, ed.
    Fernand Van Steenberghen, in *Trois commentaires anonymes sur la traité de l'âme
    d'Aristote*, pp. 121-348. Originally published as a work of Siger of Brabant in *Siger
    de Brabant d'après ses oeuvres inédites*, ed. Fernand Van Steenberghen.
    Philosophes Belges. Textes et Études 12 (Louvain, 1931).
Anonymous Vaticanus. Joachim Vennebusch, "Die Einheit der Seele nach einem anonymen
    Aristoteleskommentar aus der Zeit des Thomas von Aquin und Siger von Brabant
    (Vat. lat. 869, ff. 200r-210v)," *RTAM* 33 (1966), 37-80.
Aristotle and Averroes. *Aristotelis Opera cum Averrois Commentariis* (Venice: apud Junc-
    tas,1562-74; repr. Frankfurt a.M., 1962)
——. *Averrois Cordubensis Commentarium Magnum in Aristotelis De Anima Libros*, ed. F.
    S. Crawford (Cambridge, Mass., 1953)
Aristotle. *The Complete Works of Aristotle. The Revised Oxford Translation*, ed. Jonathan
    Barnes (Princeton, 1984).

Augustine, *De quantitate animae, PL* 32.

——, *Epist. 137 Ad Volusianum. CSEL* 44.

Avicebron. *Avicebrolis (ibn Gebirol) Fons vitae ex Arabico in Latinum translatus ab Iohanne Hispano et Dominico Gundissalino,* ed. Clemens Baeumker. *BGPM* 1, 2-4 (Münster i.W., 1892-95), pp. 1-339.

Avicenna. *Avicennae Liber De Anima seu Sextus De Naturalibus,* ed. S. van Riet (Leiden, 1968).

Boethius of Dacia. *Boetii de Dacia Tractatus de aeternitate mundi,* ed. Géza Sajó (Berlin, 1964).

Bonaventure. *Doctoris Seraphici S. Bonaventurae Opera Omnia* (10 vols., Quaracchi, 1882-1902).

——. *S. Bonaventurae Collationes in Hexaemeron et Bonaventuriana Quaedam Selecta,* ed. Ferdinand Delorme, O.F.M. (Florence/Quaracchi, 1934).

——. *Tria Opuscula Seraphici Doctoris S. Bonaventurae: Briviloquium, Itinerarium Mentis in Deum, et De reductione Artium ad Theologiam* (4th ed., Quaracchi, 1925).

Chartularium Universitatis Parisiensis, edd. Heinrich Denifle and Émile Chatelain (2 vols., Paris, 1889).

Costa ben Luca. *Excerpta e libro Alfredi Anglici De motu cordis. Item Costa-ben-Lucae De differentia animae et spiritus liber translatus a Johanne Hispalensi,* ed. C. S. Barach (Innsbruck, 1878).

Dominicus Gundissalinus. "The Treatise *De anima* of Dominicus Gundissalinus," ed. J. T. Muckle, C.S.B., with an introduction by Étienne Gilson, *Mediaeval Studies* 2 (1940), 23-102).

Giles of Rome, *Errores philosophorum,* ed. Joseph Koch, tr. John O. Riedl (Milwaukee, Wisc., 1944).

——, *Super librum de Anima, De materia celi, De intellectu possibili, De gradibus formarum* (Venice, 1500, repr. Frankfurt a. M., 1982).

John Blund. *Iohannes Blund Tractatus De Anima,* edd. D. A. Callus, O.P. and R. W. Hunt. Auctores Britannici Medii Aevi 2 (London, 1970).

John of La Rochelle. *La Summa de anima di frate Giovanni della Rochelle,* ed. Teofilo Domenichelli (Prato, 1882).

John Pecham. *Johannis Pechami Quaestiones Tractantes de Anima,* ed. P. Hieronymus Spettman, O.F.M. *BGPM* 19 (Münster i.W., 1918), 5-6.

——. *Registrum Epistolarum Fratris Johannis Peckham Archiepiscopi Cantuariensis,* ed. Charles Trice Martin. Rolls Series 3 (London, 1885, repr.1965).

——. *Tractatus de anima Ioannis Pecham,* ed. P. Gaudentius Melani, O.F.M. Biblioteca di Studi Francescani 1 (Florence, 1948).

Matthew of Aquasparta. *Matthieu d'Aquasparta, Quaestiones disputatae de anima XIII,* ed. A.-J. Gondras (Paris, 1961).

Peter of Spain. *Pedro Hispano Obras Filosóficas 1. Scientia Libri De Anima por Pedro Hispano,* ed. P. Manuel Alonso, S.J. (Madrid, 1941).

——. *Pedro Hispano Obras Filosóficas 2. Comentario al "De anima" de Aristóteles,* ed. P. Manuel Alonso, S.J. (Madrid, 1944).

——. *Pedro Hispano Obras Filosóficas 3. Expositio libri "De anima,"* ed. P. Manuel Alonso, S.J. (Madrid, 1952).

Philip the Chancellor. *Ex Summa Philippi Cancellarii Quaestiones De Anima,* ed. Leo W. Keeler, S.J. (Münster i.W., 1937).

——. *Philippi Cancellarii Parisiensis Summa de bono,* ed. Nicolas Wicki. Corpus Philosophorum Medii Aevi. Opera Philosophica Selecta (2 vols., Bern, Switzerland, 1985).

Robert Grosseteste. *Exiit edictum.* "Robert Grosseteste Bishop of Lincoln (1235-1253) On the Reasons for the Incarnation," ed. Dominic J. Unger, O.F.M.Cap., *Franciscan Studies* 16 (1956), 1-36.

———. *Robert Grosseteste De Cessatione Legalium,* edd. Richard C. Dales and Edward B. King. Auctores Britannici Medii Aevi 7 (London, 1986).

———. *Robert Grosseteste's Glosses on the Pauline Epistles,* ed. Richard C. Dales. *CCL,* continuatio medievalis (Brepols, Belgium, in press).

———. *Robert Grosseteste Hexaëmeron,* edd. Richard C. Dales and Servus Gieben, O.F.M.Cap. Auctores Britannici Medii Aevi 6 (London, 1982).

———. "Robert Grosseteste on Preaching. With an edition of the Sermon 'Ex rerum initiatarum' on Redemption," ed. Servus Gieben, O.F.M.Cap., *Collectanea Franciscana* 37 (1967), 100-41.

Robert Kilwardby. *Robert Kilwardby, O.P., De ortu scientiarum,* ed Albert G. Judy, O.P. Auctores Britannici Medii Aevi 4 (London, 1976).

———. "Le «De 43 questionibus» de Robert Kilwardby, ed. H. F. Dondaine, O.P., *AFP* 47 (1977), 5-50.

———. *Quaestiones in librum primum Sententiarum,* ed. Johannes Schneider (Munich, 1986).

———. *Quaestiones in librum tertium Sententiarum I. Christologie,* ed. Elizabeth Gössman (Munich, 1982).

———. *Quaestiones in librum tertium Sententiarum II. Tugendlehre,* ed. Gerhard Liebold (Munich, 1985).

Roger Bacon. *Opera hactenus inedita fratris Rogeri Baconis,* edd. Robert Steele and F. Delorme (16 vols., Oxford, 1905-41).

Siger of Brabant. *Les Quaestiones super librum De causis de Siger de Brabant,* ed. Antonio Marlasca. Philosophes Médiévaux 12 (Louvain/Paris, 1972).

———. *Siger de Brabant. Quaestiones in Tertium De Anima, De Anima Intellectiva, De Aeternitate Mundi,* ed. Bernardo Bazán. Philosophes Médiévaux 13 (Louvain/Paris, 1972).

Thomas Aquinas. *Sancti Thomae Aquinatis Doctoris Angelici Ordinis Praedicatorum Opera Omnia,* ed. P. Fiaccadori (Parma, 1852-73, repr. New York, 1949).

———. *De unitate intellectus. Sancti Thomae de Aquino Opera omnia,* iussu Leonis XIII P. M. edita 43 (Rome, 1976)

William of Auvergne. *Opera omnia* (Paris/Orleans, 1674; repr. Frankfurt a.M., 1963).

William of Baglione. Ignatius Brady, O.F.M., "Background to the Condemnation of 1270: Master William of Baglione, O.F.M.," *Franciscan Studies* 30(1970), 5-48.

———. *William of Baglione: Questions on the Soul (Paris, 1266-1267),* ed. Elizabeth Monica Streitz (Unpublished M.A, thesis, University of Southern California, 1991).

———. Ignatius Brady, O.F.M., "The Questions of Master William of Baglione, O.F.M., *De aeternitate mundi* (Paris, 1266-1267)," Part II, *Antonianum* 47 (1972), 576-616).

Modern Works

Anton, J., *Essays in Ancient Greek Philosophy* (Albany, N. Y., 1972).

Baeumker, Clemens, *Witelo, Ein Philosoph und Naturforscher des XIII. Jahrhunderts. BGPM* 3.2 (Münster i.W., 1908).

Bazán, Bernardo, "Le dialogue philosophique entre Siger de Brabant et Thomas d'Aquin. À propos d'un livre récent de É. H. Wéber, OP," *Revue philosophique de Louvain* 72 (1972), 53-155.

206 BIBLIOGRAPHY

—— "Pluralisme de forms ou dualisme de substances? La pensée pré-thomiste touchant la nature de l'âme," *Revue philosophique de Louvain* 67 (1969), 30-73.

Bettoni, Efrem, "Origine e struttura dell'anima umana secondo Bacone," *Rivista di Filosofia neo-scolastica* 61 (1969), 185-201.

Birkenmajer, Aleksander, "Witelo est-il l'auteur de l'opuscule "De intelligentiis'?" *Studia Copernicana* 4 (1972), 259-335.

——, *Der Brief Robert Kilwardbys an Peter von Conflans und die Streitschrift des Ägidius von Lessines*. BG-PM 20.5, (Münster i.W., 1917), pp. 36-69.

Bowman, Leonard J., "The Development of the Doctrine of the Agent Intellect in the Franciscan School of the Thirteenth Century," *The Modern Schoolman* 50 (1973), 251-79.

Callebaut, André, O.F.M., "Jean Pecham O.F.M. et l'augustinisme. Aperçus historiques (1263-1265)," *AFH* 18 (1925), 441-72.

Callus, Daniel A., O.P., "Gundissalinus' *De anima* and the Problem of Substantial Form," *The New Scholasticism* 13 (1939), 339-55.

——, "Two Early Oxford Masters on the Problem of Plurality of Forms. Adam of Buckfield—Richard Rufus of Cornwall," *RNP* 42 (1939), 411-45.

——, "Philip the Chancellor and the *De anima* ascribed to Robert Grosseteste," *Medieval and Renaissance Studies* 1 (1941), 105-27.

——, *The Condemnation of St. Thomas at Oxford* (Blackfriars [Oxford], 1946; second edition 1955).

——, "The Origins of the Problem of the Unity of Form," *The Thomist* 24 (1961), 257-88.

Chenu, M.-D., "Aux origines de la 'science moderne'," *RSPT* 29 (1940), 210-17.

——, "Les réponces de saint Thomas et de Kilwardby à la consultation de Jean de Verceil, 1271," in *Mélanges Mandonnet. Études d'histoire littéraire et doctrinale du moyen âge* (2 vols., Paris, 1930), 1, pp. 191-222.

——, Maîtres et bacheliers de l'université de Paris v. 1240," *Publications de l'Institute d'Études Médiévales d'Ottawa* (1932), 11-39.

Crowley, Theodore, *Roger Bacon. The Problem of the Soul in his Philosophical Commentaries* (Louvain /Dublin, 1950).

Da Cruz Pontes, J. M., "Le problème de l'origine de l'âme de la patristique à la solution thomistique," *RTAM* 31 (1964), 175-229.

——, *Pedro Hispano Portugalense e as controvérsias doutrinais do século XIII. A Origem da Alma* (Coimbre, 1964).

Dales, Richard C., "A Medieval View of Human Dignity," *Journal of the History of Ideas* 38 (1977), 557-72.

——, "Maimonides and Boethius of Dacia on the Eternity of the World," *The New Scholasticism* 56 (1982), 306-19.

——, *Medieval Discussions of the Eternity of the World* (Leiden, 1990).

——, "R. de Staningtona: An Unknown Writer of the Thirteenth Century," *Journal of the History of Philosophy* 4 (1966), 199-208.

——, "Robert Grosseteste's Scientific Works," *Isis* 52 (1961), 381-402.

——, "The Influence of Grosseteste's *Hexaemeron* on the *Sentences* Commentaries of Richard Rufus, O.F.M. and Richard Fishacre, O.P., *Viator* 2 (1971), 271-300.

——, "The Origins of the Doctrine of the Double Truth," *Viator* 15 (1984), 169-79.

d'Alverny, M.-T., "Dominic Gundisalvi (Gundissalinus)," in *New Catholic Encyclopedia* 4 (New York/Washington D. C., 1967), pp. 966-67.

——, "Un témoin muet des luttes doctrinales du XIIIe siècle," *AHDLMA* 17 (1949), 223-48.

Davidson, Herbert A., *Alfarabi, Avicenna, and Averroes on Intellect* (New York/Oxford, 1992).

De Rijk, L. M., "On the Life of Peter of Spain, the Author of the Tractatus Called *Summulae logicales*," *Vivarium* 8 (1970), 123-54.

de Vaux, R., "La première entrée d'Averroës chez les Latins," *RSPT* 22 (1933), 193-245.

Dondaine, A., and L. J. Bataillon, "Le manuscrit Vindob. lat. 2330 et Siger de Brabant," *AFP* 36 (1966), 153-215.

Doucet, Victorin, " A travers le manuscrit 434 de Douai," *Antonianum* 27 (1952), 531-80.

Douie, Decima, *The Conflict Between the Seculars and the Mendicants at the University of Paris in the Thirteenth Century* (New York, 1954).

Ethier, Albert-M., "La double définition de l'âme humaine chez St. Albert le Grand," *Études et Recherches publiées par le Collège dominicain d'Ottawa* I, Philosophie: Cahier I, 79-110.

Filthaut, E., *Roland von Cremona O.P. und die Anfänge der scholastik in Predigerorden. Ein Beitrag zur Geistesgechichte der älteren Dominikaner* (Vechta, 1936).

Forest, A., *La structure métaphysique du concret selon saint Thomas d'Aquin* (Paris, 1931).

Freibergs, Gunar, ed., *Aspectus et Affectus. Essays and Editions in Grosseteste and Medieval Intellectual Life in Honor of Richard C. Dales* (New York, 1993).

Gauthier, R. A., "Notes sur les débuts (1225-1240) du premier 'Averroisme'," *RSPT* 66 (1982), 322-74.

Gieben, Servus, O.F.M.Cap., "Le potenze naturali dell'anima umana secondo alcuni testi inediti di Roberto Grossatesta," *L'Homme et son destin*. Actes de premier congrès international de philosophie médiévale. (Louvain/Paris, 1960).

Gilson, Étienne, *History of Christian Philosophy in the Middle Ages* (New York, 1955).

——, "L'âme raisonnable chez Albert le Grand," *AHDLMA* 18 (1943), 5-72.

——, *La Philosophie de S. Bonaventure* (Paris, 1925).

——, "Les sources gréco-arabes de l'augustinisme avicennisant," *AHDLMA* 4 (1929-30), 5-149.

Glorieux, Palémon, "Maître Adam," *RTAM* 34 (1967), 262-70.

Goheen, John, *The Problem of Matter and Form in the* De Ente et Essentia *of Thomas Aquinas* (Cambridge, Mass., 1940).

Grabmann, Martin, "Die Aristoteles Kommentatoren Adam von Bocfeld und Adam von Bouchermefort. Die Anfänge der Erklärung des 'Neuen Aristoteles' in England," *Mittelalterliches Geistesleben* (2 vols., Munich, 1936), 2, 138-87.

——,"Mittelalterliche lateinische Aristotelesübersetzungen und Aristoteleskommentare in Handschriften spanischer Bibliotheken, "*Sitzungsberichte der Bayerischen Akademie der Wissenschaften*. Philologisch-philosophische Klasse (Munich, 1928), 98-113.

——, "Die Lehre vom Intellectus Possibilis und Intellectus Agens im Liber De Anima des Petrus Hispanus des späteren Papstes Johannes XXI," *AHDLMA* 12-13 (1937-38), 167-208.

Henquinet, F.-M., "Un brouillon autographe de S. Bonaventure sur le commentaire des Sentences," *Études Franciscaines* 44 (1932), 633-55.

Hess, C. R., "Roland of Cremona's Place in the Current of Thought," *Anglicum* 45 (1968), 420-77.

Hissette, Roland, *Enquête sur les 219 articles condamnés à Paris le 7 mars 1277*. Philosophes Médiévaux 22 (Louvain/Paris, 1977).

Hödl, Ludwig, "Anima forma corporis," *Theologie und Philosophie* 41 (1966), 536-56.

Hurley, M., "Illumination according to S. Bonaventure," *Gregorianum* 32 (1951), 388-404.

Keeler, Leo W., S.J., "The Dependence of Robert Grosseteste's *De anima* on the *Summa* of Philip the Chancellor," *The New Scholasticism* 11 (1937), 197-219.

Keicher, Otto, "Zur Lehre des ältesten Fraziskanertheologen vom 'Intellectus Agens',"
    Abhandlung aus dem Gebiete der Philosophie und ihre Geschichte (Freiburg i..B.,
    1913).

Kovach, Francis J. and Robert W. Shahan, edd., *Albert the Great. Commemorative Essays*
    (Norman, Oklahoma, 1980.

Kuksewicz, Z., "Un commentaire 'averroïste' anonyme sur le traité de l'âme d'Aristote
    (Paris, Bibl. nat. lat. 16,609, fol. 41-61)," *Revue Philosophique de Louvain* 62
    (1964), 421-65.

Long, R. James, "Richard Fishacre and the Problem of the Soul," *The Modern Schoolman* 52
    (1975), 263-70.

——— , *The Problem of the Soul in Richard Fishacre's Commentary on the Sentences*
    (Unpublished doctoral dissertation, University of Toronto, 1968).

Lottin, D. O., "La composition hylémorphique des substances spirituelles. Les debuts de la
    controverse," *RNP* 34 (1932), 21-39.

Luyckx, Bonifaz Anton, O.P., *Die Erkenntnislehre Bonaventuras. BGPM* 23, 3-4 (Münster
    i.W., 1923).

Mahoney, Edward P., "Saint Thomas and Siger of Brabant Revisited," *The Review of Meta-
    physics* 28 (1974), 531-53.

Mandonnet, Pierre, O.P., *Siger de Brabant et l'averroïsme latin au XIIIe siècle* (2 vols., 2nd
    ed., Louvain, 1911)

Marrone, Steven, *William of Auvergne and Robert Grosseteste* (Princeton, 1983).

Martin, Raymond-M., O.P., "La Question de l'Unité de la Forme substantielle dans le premier
    Collège dominicain à Oxford (1221-1248)," *RNP* 22 (1970), 107-12.

Masnovo, A., *Da Guglielmo d'Auvergne a San Tommaso* (3 vols., 2nd ed., Milan, 1945).

McEvoy, James, "Robert Grosseteste's Theory of Human Nature. With the Text of His Con-
    ference, *Ecclesia Sancta Celebrat*," *RTAM* 47 (1980), 131-87.

——— , "Grosseteste on the Soul's Care for the Body: A New Text and New Source for the
    Idea," in *Aspectus et Affectus*, ed. Gunar Freibergs (New York, 1993), 37-56.

——— , *The Philosophy of Robert Grosseteste* (Oxford, 1982).

Moody, E. A., "William of Auvergne and his Treatise *De anima*," *Studies in Medieval
    Philosophy, Science, and Logic* (Berkeley/Los Angeles, 1975), 1-109.

Nardone, Henry F., *St. Thomas Aquinas and the Condemnation of 1277.* The Catholic
    University of America Philosophical Studies No. 209 (Washington, D. C., 1963).

Nuyens, F, *L'évolution de la psychologie d'Aristote* (Louvain, 1948).

Pegis, Anton Charles, *St. Thomas and the Problem of the Soul in the Thirteenth Century* (Tor-
    onto, 1934).

Pelster, Franz, "Der älteste Sentenzenkommentar aus der Oxforder Franziskanerschule,"
    Scholastik 1 (1926), 50-80.

Prete, E., "La posizione di Rolando da Cremona nel pensiero medievale," *Rivista di Filosofia
    neoscolastica* 23 (1931), 484-89.

Raedts, Peter, *Richard Rufus of Cornwall and the Tradition of Oxford Theology* (Oxford,
    1987).

Rohmer, Jean, "Sur la doctrine franciscaine des deux faces de l'âme," *AHDLMA* 2 (1927),
    73-77.

Salman, Dominique, O.P., "Note sur la première influence d'Averroès," *RNP* 40 (1937), 205-
    12.

Southern, Richard W., *Robert Grosseteste: The Growth of an English Mind in Medieval
    Europe* (Oxford, 1986).

Szdzuj, Edouard, O.F.M., "Saint Bonaventure et la problème du rapport entre l'âme et le
    corps," *La France Franciscaine* 15 (1932), 283-310.

Thomson, S. Harrison, "The *De anima* of Robert Grosseteste," *The New Scholasticism* 7 (1933), 201-21.

——, *The Writings of Robert Grosseteste* (Cambridge, 1940).

Van Steenberghen, Fernand, *La philosophie au XIIIe siècle*. Philosophes Médiévaux 28 (2nd edition, Louvain/Paris, 1991).

Vennebusch, Joachim, "Die Einheit der Seele nach einem anonymen Aristoteleskommentar aus der Zeit des Thomas von Aquin und Siger von Brabant (Vat. lat. 869, ff. 200r-210v)," *RTAM* 33 (1966), 37-80.

Wéber, É. H., *La controverse de 1270 à l'université de Paris et son retentissement sur la pensée de S. Thomas d'Aquin*. Bibliothèque Thomiste 40 (Paris, 1979).

Weisheipl, James A., O.P., "Albertus Magnus and Universal Hylomorphism: Avicebron," in *Albert the Great. Commemorative Essays*, edd. Francis J. Kovach and Robert W. Shahan (Norman, Oklahoma, 1980), pp. 239-60.

Wilpert, Paul, "Boethius von Dacien - Die Autonomie des Philosophen," *Beiträge zum Berufsbewusstsein des mittelalterlichen Menchen* (Berlin, 1964), 135-52.

Wilshire, Leland E., "Were the Oxford Condemnations Directed Against Aquinas?" *The New Scholasticism* 48 (1974), 125-32.

——, "The Oxford Condemnations of 1277 and the Intellectual Life of the Thirteenth-Century Universities," in *Aspectus et Affectus*, ed. Gunar Freibergs (New York, 1993), 114-24.

Zavalloni, Roberto, *Richard of Mediavilla et la controverse sur la pluralité des formes. Textes inédits et étude critique*. Philosophes Médiévaux 2 (Louvain, 1951).

Zeller, E., *Die Philosophie der Griechen* (4th edition, Leipzig, 1921).

GENERAL INDEX

Abubacer 75, 114
Adam of Buckfield 47-52, 59, 65, 80, 201
Adam de Puteorum Villa (Pulchrae Mulieris) 78
Albert the Great (Saint) 27, 52, 65, 73, 74, 80, 89-98, 99, 106, 111, 117-19, 121, 134, 135, 151, 171, 190, 194, 196, 201
Alcher of Clairvaux 47, 58
Alexander of Aphrodisias 9, 32, 78, 114, 152
Alexander of Hales 27-31, 45, 47, 86, 87, 99, 102, 104, 107, 194, 197, 198, 199
Alexander Nequam 15-6, 107, 194
Alfarabi 75, 78
Alfred of Sareshel 16, 20
Algazel 120
Alonzo, Manuel 67
Anonymous Admontensis 17, 27, 78-80, 78, 84, 111, 136, 187, 195, 197
Anonymous Bazán 150-54, 181, 196
Anonymous Van Steenberghen 107, 145, 150, 163-69, 186, 195, 202
Anonymous Vaticanus 17, 27, 69, 76, 80-86, 107, 136, 187, 197, 202
Anton, J. 113n
Aristotle (the Philosopher, the Stagyrite) 1, 5, 6, 9-11, 12, 13, 15, 20, 22, 27, 28, 29, 30, 32, 33, 35, 36, 38, 45, 48, 49, 51, 58, 60, 61, 65, 72, 74, 75, 76, 77, 81, 83, 86, 89, 95, 96, 98, 99, 105n, 108, 113, 114, 118, 122, 124, 125, 126, 127, 129, 131, 133, 137, 139, 141, 142, 145, 146, 147, 148, 149, 150, 151, 152, 153, 154, 156, 157, 158, 160, 162, 164, 169, 170, 171, 172, 173, 174, 175, 180, 182, 184, 189, 193, 197, 201
Pseudo-Aristotle 65, 75
aspectus et affectus mentis 105
Augustine (Saint) 2, 3, 4, 13, 14, 16, 20, 28, 29, 35, 38, 42n, 47, 50n, 56, 58, 60 72, 83, 86, 89, 94, 100, 101, 102, 105n, 121, 126, 127, 129, 132, 133, 184, 186, 188, 191, 194, 198, 199, 202

Pseudo-Augustine 4, 47, 56, 58, 86, 199
Avempace 78, 114
Averroes (the Commentator) 3, 9n, 10n, 20, 26, 32, 36, 47, 48, 62, 65, 67, 78, 80, 86, 89, 93, 95, 106, 113-17, 118, 119, 120, 122, 124, 126, 127, 128, 131, 136, 138, 139, 142 143, 148, 149, 150, 151, 152, 153, 157, 160, 161, 162, 163, 164, 165, 166, 167, 168, 174, 176, 185, 189, 190, 193, 194, 195, 197, 201
Averroists 1, 132, 136, 140, 154, 157, 170, 181, 188, 192, 195-97
Avicebron 3, 5, 6-7, 13n, 15, 22, 32, 36, 68, 75, 78, 91, 100, 127, 128, 131, 138, 182, 199
Avicenna 5, 7-9, 13, 14, 15, 17, 20, 32, 54, 65, 68, 72, 73, 74, 77, 78, 86, 89, 95, 99, 125, 127, 130, 138, 182, 189

Baeumker, Clemens 6n, 78n
Barach, C. S. 6n
Bataillon, L. J. 176
Bazán, Bernardo 1, 3n, 32n, 132n, 150, 151, 169n
Bettoni, Efrem 75n
binarium famosissimum 14
Birkenmajer, Aleksander 63n, 78n
Boethius 3, 22, 45, 75, 78, 87, 91, 100, 111, 129, 199
Boethius of Dacia 4, 120, 133n, 134n, 148, 153, 167, 180, 181, 183, 193, 196, 197
Boethius of Dacia (?) (Anonymous Giele) 150, 154-59, 160, 163, 164, 196
Bonaventure (Saint) 57, 74, 99-107, 117, 120, 128, 138, 163, 181, 194, 195, 196, 197, 199-200
Borgnet, S. C. A. 117n
Bowman, Leonard J. 4n, 105n, 106n
Brady, Ignatius, O.F.M. 42n, 120n, 121n, 122n, 123n, 125, 184n, 190n, 202

Calcidius 86
Callebaut, André, O.F.M. 66n

BRILL'S STUDIES IN INTELLECTUAL HISTORY

1. POPKIN, R.H. *Isaac la Peyrère (1596-1676)*. 1987. ISBN 90 04 08157 7
2. THOMSON, A. *Barbary and Enlightenment.* European Attitudes towards the Maghreb in the 18th Century. 1987. ISBN 90 04 08273 5
3. DUHEM, P. *Prémices Philosophiques.* With an Introduction in English by S.L. Jaki. 1987. ISBN 90 04 08117 8
4. OUDEMANS, TH.C.W. & A.P.M.H. LARDINOIS. *Tragic Ambiguity.* Anthropology, Philosophy and Sophocles' *Antigone.* 1987. ISBN 90 04 08417 7
5. FRIEDMAN, J.B. (ed.). *John de Foxton's Liber Cosmographiae (1408).* An Edition and Codicological Study. 1988. ISBN 90 04 08528 9
6. AKKERMAN, F. & A. J. VANDERJAGT (eds.). *Rodolphus Agricola Phrisius, 1444-1485.* Proceedings of the International Conference at the University of Groningen, 28-30 October 1985. 1988. ISBN 90 04 08599 8
7. CRAIG, W.L. *The Problem of Divine Foreknowledge and Future Contingents from Aristotle to Suarez.* 1988. ISBN 90 04 08516 5
8. STROLL, M. *The Jewish Pope.* Ideology and Politics in the Papal Schism of 1130. 1987. ISBN 90 04 08590 4
9. STANESCO, M. *Jeux d'errance du chevalier médiéval.* Aspects ludiques de la fonction guerrière dans la littérature du Moyen Age flamboyant. 1988. ISBN 90 04 08684 6
10. KATZ, D. *Sabbath and Sectarianism in Seventeenth-Century England.* 1988. ISBN 90 04 08754 0
11. LERMOND, L. *The Form of Man.* Human Essence in Spinoza's *Ethic.* 1988. ISBN 90 04 08829 6
12. JONG, M. DE. *In Samuel's Image.* Early Medieval Child Oblation. (in preparation)
13. PYENSON, L. *Empire of Reason.* Exact Sciences in Indonesia, 1840-1940. 1989. ISBN 90 04 08984 5
14. CURLEY, E. & P.-F. MOREAU (eds.). *Spinoza. Issues and Directions.* The Proceedings of the Chicago Spinoza Conference. 1990. ISBN 90 04 09334 6
15. KAPLAN, Y., H. MÉCHOULAN & R.H. POPKIN (eds.). *Menasseh Ben Israel and His World.* 1989. ISBN 90 04 09114 9
16. BOS, A.P. *Cosmic and Meta-Cosmic Theology in Aristotle's Lost Dialogues.* 1989. ISBN 90 04 09155 6
17. KATZ, D.S. & J.I. ISRAEL (eds.). *Sceptics, Millenarians and Jews.* 1990. ISBN 90 04 09160 2
18. DALES, R.C. *Medieval Discussions of the Eternity of the World.* 1990. ISBN 90 04 09215 3
19. CRAIG, W.L. *Divine Foreknowledge and Human Freedom.* The Coherence of Theism: Omniscience. 1991. ISBN 90 04 09250 1
20. OTTEN, W. *The Anthropology of Johannes Scottus Eriugena.* 1991. ISBN 90 04 09302 8
21. ÅKERMAN, S. *Queen Christina of Sweden and Her Circle.* 1991. ISBN 90 04 09310 9
22. POPKIN, R.H. *The Third Force in Seventeenth-Century Thought.* 1992. ISBN 90 04 09324 9
23. DALES, R.C. & O. ARGERAMI (eds.). *Medieval Latin Texts on the Eternity of the World.* 1990. ISBN 90 04 09376 1
24. STROLL, M. *Symbols as Power.* The Papacy Following the Investiture Contest. 1991. ISBN 90 04 09374 5
25. FARAGO, C.J. *Leonardo da Vinci's 'Paragone'.* A Critical Interpretation with a New Edition of the Text in the *Codex Urbinas.* 1992. ISBN 90 04 09415 6
26. JONES, R. *Learning Arabic in Renaissance Europe.* Forthcoming. ISBN 90 04 09451 2
27. DRIJVERS, J.W. *Helena Augusta.* The Mother of Constantine the Great and the Legend of Her Finding of the True Cross. 1992. ISBN 90 04 09435 0
28. BOUCHER, W.I. *Spinoza in English.* A Bibliography from the Seventeenth-Century to the Present. 1991. ISBN 90 04 09499 7
29. McINTOSH, C. *The Rose Cross and the Age of Reason.* Eighteenth-Century Rosicrucianism in Central Europe and its Relationship to the Enlightenment.1992.ISBN 90 04 09502 0
30. CRAVEN, K. *Jonathan Swift and the Millennium of Madness.* The Information Age in Swift's *A Tale of a Tub.* 1992. ISBN 90 04 09524 1
31. BERKVENS-STEVELINCK, C., H. BOTS, P.G. HOFTIJZER & O.S. LANKHORST (eds.). *Le Magasin de l'Univers. The Dutch Republic as the Centre of the European Book Trade.* Papers from the Int. Colloquium, Wassenaar, 5-7 July 1990. 1992. ISBN 90 04 09493 8
32. GRIFFIN, JR., M.I.J. *Latitudinarianism in the Seventeenth-Century Church of England.* Annotated by R.H. Popkin. Edited by L. Freedman. 1992. ISBN 90 04 09653 1

33. WES, M.A. *Classics in Russia 1700-1855*. 1992. ISBN 90 04 09664 7
34. BULHOF, I.N. The Relationship between Literature and Science. With a Case Study in Darwin's *The Origin of Species*. 1992. ISBN 90 04 09644 2
35. LAURSEN, J.C. *The Politics of Skepticism in the Ancients, Montaigne, Hume and Kant*. 1992. ISBN 90 04 09459 8
36. COHEN, E. *The Crossroads of Justice*. Law and Culture in Late Medieval France. 1993. ISBN 90 04 09569 1
37. POPKIN, R.H. & A.J. VANDERJAGT (eds.). *Scepticism and Irreligion in the Seventeenth and Eighteenth Centuries*. 1993. ISBN 90 04 09596 9
38. MAZZOCCO, A. *Linguistic Theories in Dante and the Humanists*. 1993. ISBN 90 04 09702 3
39. KROOK, D. *John Sergeant and His Circle*. A Study of Three Seventeenth-Century English Aristotelians. Ed. with an Introduction by B.C. Southgate. 1993. ISBN 90 04 09756 2
40. AKKERMAN, F., G.C. HUISMAN & A.J. VANDERJAGT (eds.). *Wessel Gansfort (1419-1489) and Northern Humanism*. 1993. ISBN 90 04 09857 7
41. COLISH, M.L. *Peter Lombard*. 2 volumes. 1994. ISBN 90 04 09859 3 (Volume 1), ISBN 90 04 09860 7 (Volume 2), ISBN 90 04 09861 5 (Set)
42. VAN STRIEN, C.D. *British Travellers in Holland During the Stuart Period*. Edward Browne and John Locke as Tourists in the United Provinces. 1993. ISBN 90 04 09482 2
43. MACK, P. *Renaissance Argument*. Valla and Agricola in the Traditions of Rhetoric and Dialectic. 1993. ISBN 90 04 09879 8
44. DA COSTA, U. *Examination of Pharisaic Traditions*. Supplemented by SEMUEL DA SILVA's *Treatise on the Immortality of the Soul*. Translation, Notes and Introduction by H.P. Salomon & I.S.D. Sassoon. 1993. ISBN 90 04 09923 9
45. MANNS, J.W. *Reid and His French Disciples*. 1994. ISBN 90 04 09942 5
46. SPRUNGER, K.L. *Trumpets from the Tower*. English Puritan Printing in the Netherlands, 1600-1640. 1994. ISBN 90 04 09935 2
47. RUSSELL, G.A. (ed.). *The 'Arabick' Interest of the Natural Philosophers in Seventeenth-Century England*. 1994. ISBN 90 04 09888 7
48-49. SPRUIT, L. *Species intelligibilis: From Perception to Knowledge*. 2 vols. Volume I: Classical Roots and Medieval Discussions. 1994. ISBN 90 04 09883 6. Volume II: Renaissance Controversies, Later Scholasticism, and the Elimination of the Intelligible Species in Modern Philosophy. 1995. ISBN 90 04 10396 1
50. HYATTE, R. *The Arts of Friendship*. The Idealization of Friendship in Medieval and Early Renaissance Literature. 1994. ISBN 90 04 10018 0
51. CARRÉ, J. (ed.). *The Crisis of Courtesy*. Studies in the Conduct-Book in Britain, 1600-1900. 1994. ISBN 90 04 10005 9
52. BURMAN, T.E. *Religious Polemic and the Intellectual History of the Mozarabs, 1050-1200*. 1994. ISBN 90 04 09910 7
53. HORLICK, A.S. *Patricians, Professors, and Public Schools*. The Origins of Modern Educational Thought in America. 1994. ISBN 90 04 10054 7
54. MacDONALD, A.A., M. LYNCH & I.B. COWAN (eds.). *The Renaissance in Scotland*. Offered to John Durkan. 1994. ISBN 90 04 10097 0
55. VON MARTELS, Z. (ed.). *Travel Fact and Travel Fiction*. Fiction, Literary Tradition, Scholarly Discovery and Observation in Travel Writing. 1994. ISBN 90 04 10112 8
56. PRANGER, M.B. *Bernard of Clairvaux and the Shape of Monastic Thought*. Broken Dreams. 1994. ISBN 90 04 10055 5
57. VAN DEUSEN, N. *Theology and Music at the Early University*. The Case of Robert Grosseteste and Anonymous IV. 1994. ISBN 90 04 10059 8
58. WARNEKE, S. *Images of the Educational Traveller in Early Modern England*. 1994. ISBN 90 04 10126 8
59. BIETENHOLZ, P.G. *Historia and Fabula*. Myths and Legends in Historical Thought from Antiquity to the Modern Age. 1994. ISBN 90 04 10063 6
60. LAURSEN, J.C. (ed.). *New Essays on the Political Thought of the Huguenots of the Refuge*. 1995. ISBN 90 04 09986 7
61. DRIJVERS, J.W. & A.A. MacDONALD (eds.). *Centres of Learning*. Learning and Location in Pre-Modern Europe and the Near East. 1995. ISBN 90 04 10193 4
62. JAUMANN, H. *Critica*. Untersuchungen zur Geschichte der Literaturkritik zwischen Quintilian und Thomasius. 1995. ISBN 90 04 10276 0
63. HEYD, M. *"Be Sober and Reasonable."* The Critique of Enthusiasm in the Seventeenth and Early Eighteenth Century. 1995. ISBN 90 04 10118 7
64. OKENFUSS, M.J. *The Rise and Fall of Latin Humanism in Early-Modern Russia*. Pagan Authors, Ukrainians, and the Resiliency of Muscovy. 1995. ISBN 90 04 10331 7
65. DALES, R.C. *The Problem of the Rational Soul in the Thirteenth Century*. 1995. ISBN 90 04 10296 5